MODERN COLLEGE MATHEMATICS

John R. Sullivan

Clemson University

BARNES & NOBLE BOOKS
A DIVISION OF HARPER & ROW, PUBLISHERS
New York, Cambridge, Hagerstown,
Philadelphia, San Francisco, London,
Mexico City, São Paulo, Sydney

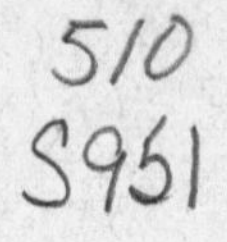

 For information address Harper & Row, Publishers, Inc., 10 East 53rd Street, New York, N.Y. 10022. Published simultaneously in Canada by Fitzhenry & Whiteside Limited, Toronto.

FIRST EDITION

Designer: Eve Kirch

Library of Congress Cataloging in Publication Data

Sullivan, John Russell.
Modern college mathematics.
(College outline series; COS/174)
Includes index.
1. Mathematics—1961- I. Title.
QA37.2.S93 510 78-15831
ISBN 0-06-460174-1 pbk.

---

80 81 82 83 84 10 9 8 7 6 5 4 3 2 1

CONTENTS

CHAPTER 3

TRIGONOMETRIC AND HYPERBOLIC FUNCTIONS

CHAPTER 6

CONTINUOUS PROBABILITY

CHAPTER 7

STATISTICAL INFERENCE

CHAPTER 8

BASIC LINEAR ALGEBRA

PREFACE

The title of this book, *Modern College Mathematics,* indicates the very broad scope of the material that is covered. Indeed, the topics that might be included in a book of this type are practically limitless. The task then becomes to select those topics that will be of greatest interest and importance to the anticipated readers and that can be accommodated within the available space.

The selection made here is confined to topics normally included in the mathematics curriculum of the first two years of college. It is assumed only that the reader has the mathematical knowledge and maturity that would be attained through basic courses in algebra and plane geometry at the high school level.

This book is designed to serve as a study guide and review of the fundamental concepts of a variety of mathematics courses. However, it is self-contained in the sense that it can be read without a background in college mathematics. In view of the widespread availability of hand-held calculators, several numerical tables that have been more or less standard in elementary mathematics texts have not been included in the appendixes.

The word "modern" in the title is indicative of such features as the use of the language and concepts of sets whenever appropriate, the stress on the concept of a function as a correspondence between two sets, and the emphasis on probability and statistical inference in accordance with that strong trend in college mathematics in recent years.

The objective of the book has been to provide an overview of a large number of topics, hence the presentation of many of them is necessarily brief.

John R. Sullivan

1

BASIC ALGEBRA OF NUMBERS AND SETS

1.1 Definitions and Notation of Sets

A **set** is any well-defined collection of objects. The objects may be numbers, persons, equations, TV sets, any entities whatsoever. Each member of the collection is called an **element** of the set. The number of elements in any set may be finite or infinite. For example, the group of all former presidents of the United States, living and dead, is a finite set, whereas the collection of all positive integers constitutes an infinite set. Note that the requirement that a set be well-defined is satisfied in these two examples. There is no question as to which individuals were presidents or which integers are positive. However, the collection of all attractive houses in the state of South Carolina would not constitute a set since there would be differing opinions as to whether a particular house belonged to the set or not.

A set is often specified by enclosing its elements in braces. For example, the set of positive integers less than or equal to 5 could be exhibited as $\{1,2,3,4,5\}$. The set of positive integers could be shown as $\{1,2,3, \ldots ,\}$, where the three dots indicate that the elements of the set continue without end.

Another notation in common use is illustrated by the set $\{x|x$ is an integer between 5 and 10 inclusive$\}$. In this case the vertical line stands for the words "such that," and the set may be described in words as the set of numbers x such that x is an integer between 5 and 10 inclusive. This set is, of course, identical with the set $\{5,6,7,8,9,10\}$.

Sets are often specified by capital letters. If a is an element of the set A, this fact is indicated by the notation $a \in A$. The notation $a \notin A$ is used to indicate that a is *not* an element of A.

A set A is said to be a **subset** of a set B if each element of A is also an element of B. The notation used to indicate this fact is $A \subset B$. This concept leads to a definition of equality for sets. Two sets are said to be equal, i.e., $A = B$, if and only if $A \subset B$ and $B \subset A$. It is therefore clear that two sets are equal if and only if they contain exactly the same elements. Finally, A is said to be a proper subset of B if and only if $A \subset B$ but $A \neq B$. In other words, A is a proper subset of B only when B contains not only each of the elements of A but also one or more additional elements. For example, the set $A = \{1,3,5,7\}$ is a proper subset of the set $B = \{1,3,5,7,9,11\}$.

Just as we have a zero in our system of real numbers, we find it necessary to define the set that contains no elements. This, called the **empty** set, is often indicated by the symbol $\varnothing$.

The **universal set,** usually indicated by the letter U, consists of the entire set of elements under consideration at a given time. For example, in a given discussion the universal set might consist of the set of all real numbers.

The **complement of a set** A consists of all of the elements of U that are not contained in A. It is commonly denoted by the symbol A'.

A concept somewhat similar to that of the complement of a set is that of the **difference of two sets.** The difference of two sets A and B, denoted by the symbol $A - B$, consists of all of the elements in A that are not contained in B. Evidently, $U - A = A'$.

Two sets are said to be disjoint if they have no element in common. Thus two sets A and B are disjoint provided that if $a \in A$, then $a \notin B$, where a is any element of A.

A convenient pictorial representation of sets is provided by **Venn diagrams.** These consist of circles that represents sets. They are

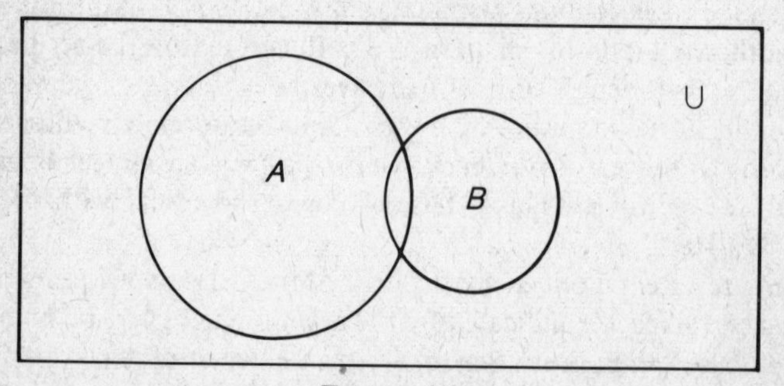

FIGURE 1.1

often shown inside a rectangle that represents the universal set. A typical Venn diagram is shown in Figure 1.1.

As will be seen, Venn diagrams, combined with appropriate shading, provide useful representations of the operations of set algebra and can conveniently be employed in the proofs of simple theorems.

1.2 The Real Numbers

We are, of course, familiar with the numbers that are called "real" since we have used them throughout our schooling. The set of real numbers may be divided into two subsets, the rational numbers and the irrational numbers. A **rational number** is defined as a number that can be expressed as the quotient of two integers, i.e., two whole numbers. It is readily seen that the rational numbers consist of the integers and the fractions. Note that an integer such as 5 can be expressed as $5/1$. Any real number that is not rational, i.e., any number that cannot be expressed as the quotient of two integers, is called **irrational.** Examples of irrational numbers are the square root of 2 and the number π, which is the ratio of the circumference of a circle to its diameter. The proof that $\sqrt{2}$ is irrational involves assuming that $\sqrt{2} = p/q$, where p and q are integers, and showing that this assumption leads to a contradiction.

The set of real numbers can be conveniently represented geometrically. Consider the straight line, assumed to be infinite in extent, shown in Figure 1.2.

Choose any point on the line and call it the **origin.** Let the origin correspond to the number 0. Then choose a unit of length and lay off a succession of these lengths both to the right and to the left of the origin. Let the points thus obtained to the right of the origin correspond to the positive integers, and let those to the left of the origin correspond to the negative integers, as shown in the figure. Such a line will be referred to as a **real number line.** Clearly the points corresponding to the fractions will lie between the points corresponding to the integers. For example, the point corresponding to $1/3$ lies one-third of the way from the point labeled 0 to the point labeled 1 in the figure. Moreover, the irrational numbers are also represented by points on the line. In fact, it can be shown

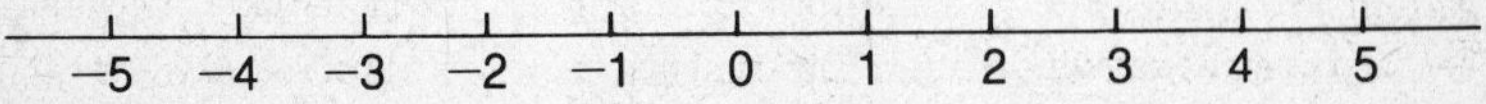

−5 −4 −3 −2 −1 0 1 2 3 4 5

FIGURE 1.2

that to each element in the set of real numbers there corresponds one and only one element of the set of points on a real number line. This is known as a **one-to-one correspondence.**

The geometrical representation of real numbers provides a useful basis for defining the concepts of **greater than** and **less than.** A number a is said to be greater than a number b if it lies to the right of b on the real number line. This situation is represented symbolically by the notation $a > b$. Similarly, a number a is said to be less than a number b if a lies to the left of b on the real number line. The symbolic representation in this case is $a < b$. Sometimes the equality and the inequality symbols are used jointly. Thus $a \geq b$ stands for *a is greater than or equal to b,* and $a \leq b$ stands for *a is less than or equal to b.*

Using the inequality symbols, we can formulate alternative but equivalent definitions of *greater than* and *less than* as follows: $a > b$ if and only if $a - b > 0$ and $a < b$ if and only if $a - b < 0$.

A useful notation that involves inequalities but that does not make use of the inequality symbols is called **interval notation.** The symbol $[a,b]$ represents the set $\{x|a \leq x \leq b\}$ and is called a **closed interval.** The symbol (a,b) represents the set $\{x|a < x < b\}$ and is called an **open interval.** In other words, a closed interval includes its endpoints, whereas an open interval does not. If we wish to designate an interval that includes one endpoint but not the other (sometimes called a half-open interval), then we may use the symbols $[a,b)$ or $(a,b]$, where $[a,b) = \{x|a \leq x < b\}$ and $(a,b] = \{x|a < x \leq b\}$.

Finally, by use of the symbol ∞ (read as infinity) the following interval notation is commonly used:

$$(-\infty,b) = \{x|x < b\}$$
$$(-\infty,b] = \{x|x \leq b\}$$
$$(a,\infty) = \{x|x > a\}$$
$$[a,\infty) = \{x|x \geq a\}$$
$$(-\infty,\infty) = \{x|x \text{ is a real number}\}$$

Irrational numbers are often approximated by rational numbers. For example, $\sqrt{2}$ approximately equals (symbolically $\simeq$) 1.414, correct to three decimal places, and 1.414 is a rational number since it can be expressed as 1414/1000. This example makes it clear that any terminating decimal number is a rational number. However, not all rational numbers correspond to terminating decimals. For example, $2/3 = 0.6666...$, where the three dots indicates that the succession is endless.

A concept of importance in connection with real numbers is that of the **absolute value** of a number. The absolute value of a number x is indicated by the symbol $|x|$, and it may be defined as follows:

$$|x| = x, \qquad \text{if } x \geq 0$$
$$|x| = -x \qquad \text{if } x < 0$$

For example, $|7| = 7$ and $|-3| = 3$. Note that the absolute value of a number is always non-negative whether the number itself is positive or negative. An important and useful inequality involving absolute values is as follows:

$$|a + b| \leq |a| + |b|$$

If you try some examples, you will see that the inequality symbol applies when a and b are of opposite signs and that the equality symbol holds when they have the same sign.

1.3 The Rectangular Coordinate System

The representation of real numbers as points on a line can readily be extended to two dimensions to give rise to what is known as the **rectangular coordinate system,** which provides the basis for the subject of analytic geometry, i.e., the study of geometry by algebraic methods.

To establish a rectangular coordinate system we draw two mutually perpendicular real number lines. See Figure 1.3. Their point of intersection is taken as the point corresponding to 0 on each of the lines. This point is called the **origin.** The lines are called the **horizontal** and **vertical axes** of the system. They are also referred to collectively as the **coordinate axes.**

Note that the horizontal line is identical with the line in Figure 1.2, with the positive numbers appearing to the right and the negative numbers to the left of the origin. On the vertical line the positive numbers correspond to points above the origin and the negative numbers to points below it.

We are now in a position to define the concept of an ordered pair and to establish a correspondence between ordered pairs and points in the rectangular coordinate plane. An **ordered pair** is simply a set of two real numbers for which the order of appearance is significant. They are usually written surrounded by parentheses and

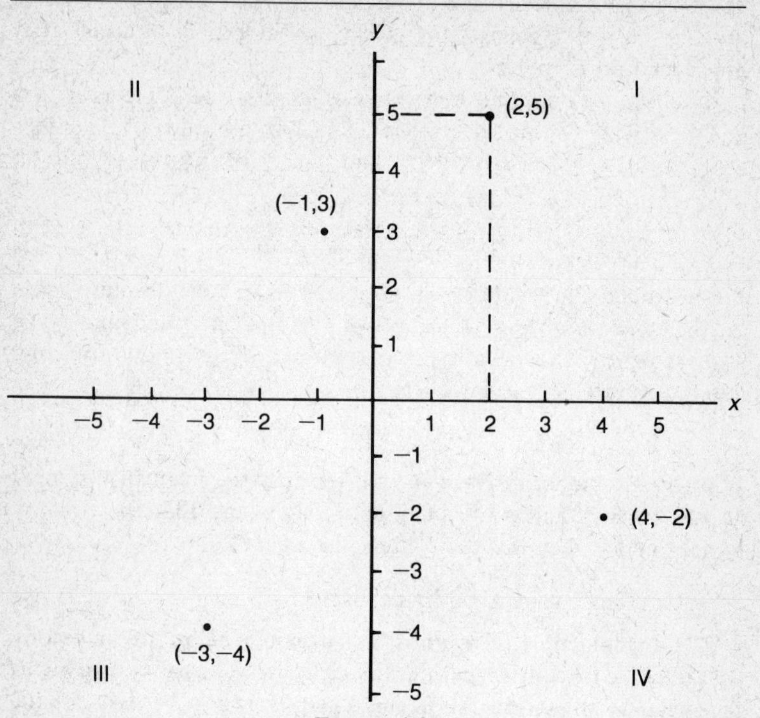

FIGURE 1.3

separated by a comma. For example, the symbols (2,5) and (5,2) represent two distinct ordered pairs, employing the same numbers but in reverse order. Unfortunately, the notation for an ordered pair is identical with that for an open interval, but the meaning will always be clear from the context.

To establish the correspondence between the ordered pair (2,5) and a point in the rectangular coordinate plane proceed as follows. Draw a vertical line through the point corresponding to 2 on the horizontal axis and a horizontal line through the point corresponding to 5 on the vertical axis. The point of intersection of these two lines, shown as broken lines in Figure 1.3, corresponds to the ordered pair (2,5).

The coordinate axes divide the plane into four parts commonly called **quadrants.** The first quadrant consists of the points above the horizontal and to the right of the vertical axis, and the other quadrants are numbered by proceeding in the counterclockwise direction. The quadrants are numbered in Figure 1.3, with Roman numerals. Ordered pairs containing only positive numbers correspond to points in the first quadrant. Pairs in which the first number is negative and the second is positive correspond to points in the second quadrant, etc. Various examples are shown in Figure 1.3. Also, all pairs whose first number is 0 correspond to points on the vertical axis, and pairs whose second number is 0 correspond to points on the horizontal axis. Points lying on the coordinate axes are not said to be in any quadrant.

It can be shown that there exists a one-to-one correspondence between the set of all points in the rectangular coordinate plane and the set of all ordered pairs of real numbers. The numbers of an ordered pair are called the **coordinates** of the corresponding point. The symbol (x,y) is used as a general designation of an ordered pair, and this leads to the designation of the horizontal axis as the x axis and the vertical axis as the y axis of the rectangular coordinate system. The first coordinate of a point is often called the x coordinate and the second coordinate the y coordinate. These are also referred to as the **abscissa** and the **ordinate,** respectively.

As stated above, the rectangular coordinate system serves as the basis for the algebraic study of geometry, a subject that will be explored in Chapter 2. It also provides the framework for the definition of the trigonometric functions, which will be discussed in Chapter 3.

1.4 Algebraic Laws for Real Numbers

The algebraic operations that will be used in this and succeeding chapters are based on the four laws listed below. The symbols used are assumed to represent real numbers.

(1) Commutative law for addition
$a+b=b+a$

(2) Commutative law for multiplication
$a \cdot b=b \cdot a$

(3) Associative law for addition
$(a+b)+c=a+(b+c)$

(4) Associative law for multiplication
$(ab)(c)=(a)(bc)$

(5) Distributive law
$a(b+c)=ab+ac$

1.5 Operations of Set Algebra

The two basic operations of set algebra are those of union and intersection. The **union** of two sets A and B is the set that contains each of the elements contained in either A or in B or in both. The union is indicated by the symbol $A \cup B$. The **intersection** of two sets A and B is the set that consists of each element contained in both A and B. The intersection is indicated by the symbol $A \cap B$. Clearly $(A \cap B) \subset (A \cup B)$.

The concepts of union and intersection can be conveniently illustrated using shaded areas in Venn diagrams as shown in Figure 1.4.

For example, suppose $A = \{1,2,4,7,9\}$ and $B = \{2,4,8,10\}$. Then $A \cup B = \{1,2,4,7,8,9,10\}$ and $A \cap B = \{2,4\}$. Clearly, $A \cap B \subset A \cup B$, for $\{2,4\} \subset \{1,2,4,7,8,9,10\}$.

The ideas of union and intersection can readily be extended to cases involving more than two sets. Thus $A \cup B \cup C$ stands for the sets of elements contained in either A, B, or C or in any combination of two or more of the three sets. $A \cap B \cap C$ stands for the set of elements that are contained in all three of the sets A, B, and C.

Another useful operation is taking the Cartesian product of two sets. The **Cartesian product** of two sets A and B is defined as the set of all possible ordered pairs in which the first element is an element of set A and the second element is an element of set B. The Cartesian product is denoted by the symbol $A \times B$. In accordance with the definition, the product $A \times B$ is not in general equal to the product $B \times A$.

For example, suppose $A = \{1,4,8\}$ and $B = \{2,4,7,9\}$. Then

$$A \times B = \{(1,2),(1,4),(1,7),(1,9),(4,2),(4,4),(4,7),(4,9),(8,2),(8,4),(8,7),(8,9)\}$$

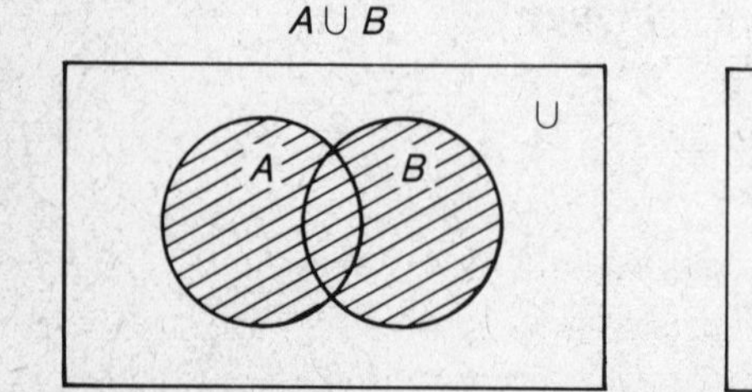

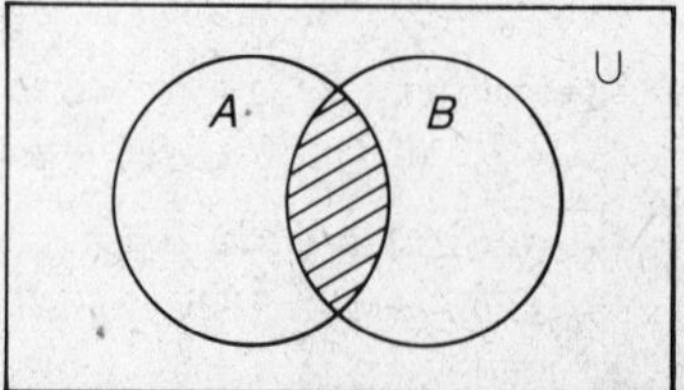

FIGURE 1.4

whereas

$$B \times A = \{(2,1),(2,4),(2,8),(4,1),(4,4),(4,8),(7,1),(7,4),(7,8),(9,1),(9,4),(9,8)\}$$

Of course we can take the cross-product of a set with itself. For example, consider the set of all possible outcomes of the experiment of throwing a single die. If this set is represented by the letter C, we have $C = \{1,2,3,4,5,6\}$. The set $C \times C$, which will have 36 elements and will be a representation of all the possible outcomes of the experiment of throwing two dice.

1.6 Laws of Set Algebra

The algebra of sets may be shown to obey a series of basic laws similar to those of the algebra of real numbers. Some of these are listed below.

(1) The commutative law for union
$A \cup B = B \cup A$

(2) The commutative law for intersection
$A \cap B = B \cap A$

(3) The associative law for union
$(A \cup B) \cup C = A \cup (B \cup C)$

(4) The associative law for intersection
$(A \cap B) \cap C = A \cap (B \cap C)$

(5) The distributive law for union over intersection
$A \cup (B \cap C) = (A \cup B) \cap (A \cup C)$

(6) The distributive law for intersection over union
$A \cap (B \cup C) = (A \cap B) \cup (A \cap C)$

Two additional laws that are of considerable interest and importance are the two so-called De Morgan laws:

(7) $(A \cup B)' = A' \cap B'$

(8) $(A \cap B)' = A' \cup B'$

These laws have important applications in logic.

Each of the above laws can be proven, some of the proofs being rather obvious. For example, (1) and (2) are evident from the definitions of union and intersection. In some cases, Venn diagrams are

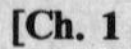

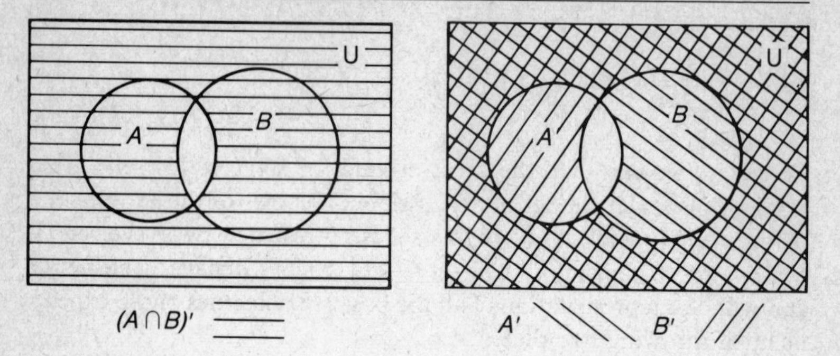

FIGURE 1.5

useful in formulating proofs. In Figure 1.5 a Venn diagram is used to prove (8).

The entire shaded area in the right-hand diagram represents $A' \cup B'$ and is seen to correspond to the shaded area in the left-hand diagram that represents $(A \cap B)'$.

In connection with the associative laws it should be noted that although we can manipulate parentheses rather freely when only union or intersection is involved, we cannot do so when a combination of the two operations is involved. For example,

$$A \cup (B \cap C) \neq (A \cup B) \cap C$$

The truth of this statement may be shown by an example.

Let $A = \{1,2,4\}$, $B = \{2,5,7\}$, $C = \{1,5,7\}$. Then $B \cap C = \{5,7\}$ and $A \cup (B \cap C) = \{1,2,4\} \cup \{5,7\} = \{1,2,4,5,7)\}$. On the other hand, $A \cup B = \{1,2,4,5,7\}$ and $(A \cup B) \cap C = \{1,2,4,5,7\} \cap \{1,5,7\} = \{1,5,7\}$.

1.7 Exponents and Radicals

We assume initially that the **exponents** we discuss are positive integers. Such exponents provide a shorthand notation for writing an extended product in which the factors are identical. Thus if a is any real number, then

$$a^5 = a \cdot a \cdot a \cdot a \cdot a$$

where clearly the exponent 5 corresponds to the number of times the factor a appears in the product. We can then state the following basic laws for operating with positive integral exponents:

(1) $a^p \cdot a^q = a^{p+q}$

(2)* $\dfrac{a^p}{a^q} = a^{p-q} \qquad p > q,\ a \neq 0$†

(3) $\dfrac{a^p}{a^q} = \dfrac{1}{a^{q-p}} \qquad p < q,\ a \neq 0$

(4) $(a^p)^q = a^{pq}$

(5) $(ab)^p = a^p \cdot b^p$

(6) $\left(\dfrac{a}{b}\right)^p = \dfrac{a^p}{b^p} \qquad b \neq 0$

In connection with these laws, the symbol a^p is commonly read as "a to the pth power."

Example 1:

$$a^3 \cdot a^5 = a^8$$

Example 2:

$$\frac{a^7}{a^3} = a^4$$

Example 3:

$$\frac{a^2}{a^5} = \frac{1}{a^3}$$

Example 4:

$$(a^2)^3 = a^6$$

Example 5:

$$(ab)^4 = a^4 b^4$$

Example 6:

$$\left(\frac{a}{b}\right)^6 = \frac{a^6}{b^6}$$

The laws of exponents can readily be extended to include negative, fractional, and zero exponents provided that these are properly defined.

* Evidently $a^p/a^q = 1$ if $p = q$.
† The symbol $\neq$ means "is not equal to."

To define the zero exponent, note that if we extend law (2) to apply to the case in which $q = p$, we have $a^p/a^p = a^{p-p} = a^0 = 1$. Hence we define the **zero exponent** such that any real number except 0 raised to the zero power is equal to 1. The expression 0^0 is not defined.

To define **negative exponents,** note that in accordance with law (3), $a^3/a^5 = 1/a^2$, but that law (2) implies $a^3/a^5 = a^{-2}$. Hence it seems reasonable to define a^{-p} as being equal to $1/a^p$, provided $a \neq 0$. Once this has been done, law (3) of course becomes superfluous.

We now consider exponents of the form p/q, where p and q are positive integers. In so doing, we shall establish a relationship between exponents and **radicals.**

The number b is said to be a qth root of the number a if and only if $b^q = a$. For example, 4 is a square root of 16 and -3 is a cube root of -27. In those cases in which a given number has more than one real qth root, only one of these is non-negative. In other cases there is only one real qth root, which may be negative or non-negative. In still other cases there are no real roots whatever. For example, 25 has two real square roots, -5 and 5; -64 has one real cube root, -4; and -36 has no real fourth roots.

The expression $\sqrt[q]{a}$, where a is a real number, is called a **radical expression.** The symbol $\sqrt[q]{a}$ is called a **radical,** and the number a is called the **radicand.**

The expression $\sqrt[q]{a}$ is defined to be the **principal qth root** of a in the following sense.

(1) $\sqrt[q]{a}$ stands for the non-negative real qth root of a if there is such a root.

(2) Otherwise $\sqrt[q]{a}$ stands for the real qth root of a if there is such a root.

For example, $\sqrt{4} = 2$ and $-\sqrt{4} = -2$. Also, $\sqrt[3]{-8} = -2$. The number -8 has two other cube roots, but they are not real numbers. Finally, $\sqrt[4]{-16}$ has no meaning in the context of real numbers. It will be assigned a meaning in Section 1.10.

Returning to exponents, in accordance with law (4), we have $(a^{1/q})^q = a^{q/q} = a$. Hence $a^{1/q}$ is a qth root of a, and we define it to be the principal qth root of a; in other words,

$$a^{1/q} = \sqrt[q]{a}$$

Furthermore

$$a^{p/q} = (a^{1/q})^p = (a^p)^{1/q}$$

so that

$$a^{p/q} = (\sqrt[q]{a})^p = \sqrt[q]{a^p}$$

Hence, in order to evaluate $a^{p/q}$ we find the principal qth root of a and raise it to the pth power, or we may first raise a to the pth power and then find the principal qth root of the resulting number.

Example 7:

$$5^0 = 1$$

Example 8:

$$3^{-2} = 1/3^2 = 1/9$$

Example 9:

$$4^{3/2} = \sqrt{4^3} = \sqrt{64} = 8$$

Example 10:

$$64^{2/3} = (\sqrt[3]{64})^2 = 4^2 = 16$$

It should be noted that we have now assigned a meaning to any exponent that is a rational number. Furthermore, it may be shown that the laws of exponents, previously stated for positive integral exponents, are valid for all rational exponents, i.e., for all exponents of the form p/q, where p and q are integers and $q \neq 0$. For example, $3^{2/3} \cdot 3^{-1/5} = 3^{2/3-1/5} = 3^{7/15}$ and $(3^{2/3})^{-1/5} = 3^{-2/15}$.

The basic laws of operation with radicals may be deduced from those of exponents, since radical expressions may be written in exponential form. They may be stated as shown below. It is assumed that all of the roots involved are real numbers.

(1) $\sqrt[p]{a} \cdot \sqrt[p]{b} = \sqrt[p]{ab}$

(2) $\dfrac{\sqrt[p]{a}}{\sqrt[p]{b}} = \sqrt[p]{\dfrac{a}{b}}$

(3) $\sqrt[p]{\sqrt[q]{a}} = \sqrt[pq]{a}$

Example 11:

$$\sqrt[3]{2} \cdot \sqrt[3]{4} = \sqrt[3]{8} = 2$$

Example 12:

$$\frac{\sqrt[4]{48}}{\sqrt[4]{3}} = \sqrt[4]{16} = 2$$

Example 13:

$$\sqrt[4]{\sqrt[3]{5}} = \sqrt[12]{5}$$

As indicated above, these laws can be easily proved by writing each of the expressions involved in exponential form and applying the laws of exponents.

Sometimes a radical expression can be simplified by factoring the radicand and then applying law (1). Some examples follow.

Example 14:

$$\sqrt{48} = \sqrt{(3) \cdot (16)} = \sqrt{3} \cdot \sqrt{16} = 4\sqrt{3}$$

Example 15:

$$\sqrt[3]{56} = \sqrt[3]{(7) \cdot (8)} = \sqrt[3]{7} \cdot \sqrt[3]{8} = 2\sqrt[3]{7}$$

Example 16:

$$\sqrt{x^3} = \sqrt{x^2 \cdot x} = x\sqrt{x} \qquad \text{provided } x \geq 0$$

The reason for the restriction $x \geq 0$ in Example 16 is that if $x \geq 0$, then $\sqrt{x^2} = x$; but if $x < 0$, then $\sqrt{x^2} = -x$. For example, if $x = 3$, then $\sqrt{x^2} = \sqrt{9} = 3 = x$; but if $x = -3$, then $\sqrt{x^2} = \sqrt{9} = 3 = -x$. In summary, it can be stated that $\sqrt{x^2} = |x|$.

Example 17:

$$\begin{aligned} \sqrt[4]{x^5y^7} &= \sqrt[4]{x^4y^4 \cdot xy^3} \\ &= \sqrt[4]{x^4y^4} \cdot \sqrt[4]{xy^3} = xy\sqrt[4]{xy^3} \end{aligned}$$

provided x and y have the same sign.

A fraction containing a radical expression in the denominator may be simplified by a process called **rationalizing** the denominator. For example, the expression $2/\sqrt{3}$ may be simplified as follows:

$$\frac{2}{\sqrt{3}} = \frac{2}{\sqrt{3}} \cdot \frac{\sqrt{3}}{\sqrt{3}} = \frac{2\sqrt{3}}{\sqrt{9}} = \frac{2\sqrt{3}}{3}$$

The last expression is simpler than the original one in the sense that it is easier to calculate if a decimal approximation is required. To calculate $2/\sqrt{3}$, we must divide 2 by $\sqrt{3}$ ($\simeq 1.732$). This is a long division problem. On the other hand, to calculate $2\sqrt{3}/3$, we multiply 1.732 by 2 and divide by 3, and only short division is involved.

The process illustrated above may be applied to more complicated expressions:

$$\frac{3}{2-\sqrt{5}} = \frac{3}{2-\sqrt{5}} \cdot \frac{2+\sqrt{5}}{2+\sqrt{5}} = \frac{6+3\sqrt{5}}{4-5} = -6-3\sqrt{5}$$

$$\frac{5}{\sqrt{2}+\sqrt{7}} = \frac{5}{\sqrt{2}+\sqrt{7}} \cdot \frac{\sqrt{2}-\sqrt{7}}{\sqrt{2}-\sqrt{7}} = \frac{5(\sqrt{2}-\sqrt{7})}{2-7} = \sqrt{7}-\sqrt{2}$$

Finally, radical expressions involving different roots may be multiplied or divided by changing over to exponential form.

Example 18:

$$\begin{aligned} \sqrt{a} \cdot \sqrt[3]{b} &= a^{1/2} \cdot b^{1/3} \\ &= a^{3/6} \cdot b^{2/6} = (a^3)^{1/6} \cdot (b^2)^{1/6} \\ &= \sqrt[6]{a^3} \cdot \sqrt[6]{b^2} \\ &= \sqrt[6]{a^3 b^2} \end{aligned}$$

1.8 Factoring

We now proceed to a brief review of a well-known algebraic process called factoring. **Factoring** may be described as reverse multiplication. For example, the multiplication formula $(a+b) \cdot (a-b) = a^2 - b^2$ becomes, when read in reverse order, the well-known formula for factoring the difference between two squares.

The simplest and most commonly used factoring formula is the one that concerns the removal of a common factor from a sum of several terms, for example, $ab + ac + ad = a(b + c + d)$. This is of course simply the reversal of the distributive law.

Two other common factoring formulas are those concerning the sum or difference of two cubes. These are

$$a^3 + b^3 = (a+b)(a^2 - ab + b^2)$$

$$a^3 - b^3 = (a-b)(a^2 + ab + b^2)$$

These can, of course, be readily verified by multiplication.

Certain expressions can be factored by first regrouping the terms and then removing common factors from each of the groups. For example,

$$ax + by + bx + ay = ax + bx + ay + by$$
$$= x(a + b) + y(a + b) = (a + b)(x + y)$$

In some cases, the formulas shown above can be used repeatedly or in combination. Here are some examples.

Example 1:

$$\begin{aligned}(x^4 - y^4) &= (x^2)^2 - (y^2)^2 \\ &= (x^2 + y^2)(x^2 - y^2) \\ &= (x^2 + y^2)(x + y)(x - y).\end{aligned}$$

Example 2:

$$\begin{aligned}(x^6 - y^6) &= (x^3)^2 - (y^3)^2 \\ &= (x^3 + y^3)(x^3 - y^3) \\ &= (x + y)(x^2 - xy + y^2)(x - y)(x^2 + xy + y^2)\end{aligned}$$

Example 3:

$$\begin{aligned}x^9 + y^9 &= (x^3)^3 + (y^3)^3 \\ &= (x^3 + y^3)(x^6 - x^3y^3 + y^6) \\ &= (x + y)(x^2 - xy + y^2)(x^6 - x^3y^3 + y^6)\end{aligned}$$

Example 4:

$$\begin{aligned}ax^2 - 4by^2 - 4ay^2 + bx^2 &= ax^2 - 4ay^2 + bx^2 - 4by^2 \\ &= a(x^2 - 4y^2) + b(x^2 - 4y^2) \\ &= (x^2 - 4y^2)(a + b) \\ &= (x + 2y)(x - 2y)(a + b)\end{aligned}$$

1.9 Variation

The subject of variation involves simple cases of a mathematical entity called a function, which will be discussed in detail in Chapter 2.

A quantity y is said to **vary directly** with a quantity x, or to be **directly proportional** to x, if there is some constant k such that the relationship $y = kx$ holds for all values of x and y. A quantity y is said to **vary inversely** with x, or to be **inversely proportional** to x, if the relationship $y = k/x$ holds for all values of x and y. Note that in the case of direct variation, x and y increase or decrease together, whereas in the case of inverse variation, an increase in x produces a decrease in y and vice-versa. Direct and inverse variation have many applications. For example, a student's numerical grade in a course may be assumed to vary directly with the number of hours of study, whereas the amount of money in the student's wallet at the end of a given week will vary inversely with the number of movies attended during the week.

Example 1 A quantity y varies inversely with a quantity x. It has been observed that when $x = 4$, $y = 16$. Find the law of variation.

Solution Since the variation is inverse, we have $y = k/x$. Substituting the observed values gives $16 = k/4$. Solving for k by multiplying both sides of the equation by 4, we obtain $k = 64$. Therefore the law of variation is $y = 64/x$.

Example 2 Newton's law of universal gravitation states that the force of attraction between any two particles is directly proportional to the product of their masses and inversely proportional to the square of the distance between them. Express this law by means of an equation.

Solution If the masses are denoted by m_1 and m_2, the distance between them by d, and the force by F, then $F = km_1m_2/d^2$.

1.10 Imaginary and Complex Numbers

The real numbers are actually a subset of a set of numbers called **complex.** The need for numbers other than real numbers becomes apparent when we attempt to solve an equation such as $x^2 + 1 = 0$. Evidently this equation is equivalent to the equation $x^2 = -1$. It is clear from the laws of signs for multiplication that there is no real number whose square is -1. We are thus led to the invention of a new kind of number, which is called an **imaginary number.** This term, although traditional, is somewhat unfortunate since it suggests that such numbers do not really exist, whereas they are simply numbers of a different type, and they have important applications in both pure and applied mathematics.

We begin by defining a number, to be denoted by the letter i, such that $i = \sqrt{-1}$. Hence, $i^2 = -1$. Furthermore, assuming that the laws of exponents for real numbers hold for our new number, we have

$$i^3 = i^2 \cdot i = (-1)(i) = -i$$
$$i^4 = (i^2)^2 = (-1)^2 = 1$$
$$i^5 = (i^4)i = (1)(i) = i$$

Evidently, if we continue in this way to obtain higher powers of i, the values i, -1, $-i$, 1 will continue to recur in that order. Also, if we assume that the law for multiplication of radical expressions holds for $\sqrt{(a^2)(-1)}$, where a is a real number, we have, provided $a \geq 0$,

$$\sqrt{-a^2} = \sqrt{(a^2)(-1)} = \sqrt{a^2} \cdot \sqrt{-1} = ai$$

For example,

$$\sqrt{-9} = 3i \text{ and } \sqrt{-8} = 2\sqrt{2}\,i$$

A number of the form $a + bi$, where a and b are real numbers, is called a **complex number.** The number a is called the **real part** of the complex number and the number b is called the **imaginary part.**

The subset of the set of complex numbers for which $b = 0$ constitutes the set of real numbers, or $R = \{a + bi | b = 0\}$. The subset of the set of complex numbers for which $a = 0$ and $b \neq 0$ is called the set of pure imaginary numbers. All complex numbers for which $b \neq 0$ are called imaginary numbers.

Two complex numbers $a + bi$ and $c + di$ are said to be equal if and only if $a = c$ and $b = d$, i.e., if and only if their real parts are equal and their imaginary parts are equal.

The fundamental algebraic operations are defined for complex numbers as follows:

(1) $(a + bi) + (c + di) = (a + c) + (b + d)i$

(2) $(a + bi) - (c + di) = (a - c) + (b - d)i$

(3) $(a + bi) \cdot (c + di) = (ac - bd) + (ad + bc)i$

(4) $\dfrac{a + bi}{c + di} = \dfrac{ac + bd}{c^2 + d^2} + \left(\dfrac{bc - ad}{c^2 + d^2}\right)i$

These definitions are motivated by our desire that when algebraic operations are applied to complex numbers the commutative, associative, and distributive laws are to be retained.

For example, if we multiply $a + bi$ and $c + di$ in the usual way, as though all of the numbers were real, we obtain

$$\begin{aligned}(a + bi) \cdot (c + di) &= ac + (ad + bc)i + bdi^2 \\ &= (ac - bd) + (ad + bc)i \qquad \text{since } i^2 = -1\end{aligned}$$

We note that this agrees with definition (2) above.

In dividing $a + bi$ by $c + di$ we make use of the notion of the **conjugate** of a complex number. The numbers $a + bi$ and $a - bi$ are said to be conjugate complex numbers; either number is said to be the conjugate of the other. Furthermore, it is readily seen that the product of two conjugate complex numbers is a real number since

$$(a + bi) \cdot (a - bi) = a^2 - b^2i^2 = a^2 + b^2$$

This fact is employed in the division of complex numbers as follows:

$$\begin{aligned}\frac{a+bi}{c+di} &= \frac{a+bi}{c+di} \cdot \frac{c-di}{c-di} \\ &= \frac{(ac+bd)+(bc-ad)i}{c^2+d^2} \\ &= \frac{ac+bd}{c^2+d^2} + \left(\frac{bc-ad}{c^2+d^2}\right)i\end{aligned}$$

This agrees with definition (3) above.

In view of the above, the student, in performing operations on complex numbers, may, instead of applying definitions (1) through (4) directly, simply carry out the operations as though only real numbers were involved, replacing i^2 wherever it occurs by -1:

Example 1:

$$(2 + 3i) + (5 - 7i) = 7 - 4i$$

Example 2:

$$(2 + 3i) \cdot (5 - 7i) = 10 + i - 21i^2 = 31 + i$$

Example 3:

$$\begin{aligned}\frac{2+3i}{5-7i} &= \frac{2+3i}{5-7i} \cdot \frac{5+7i}{5+7i} = \frac{10+29i+21i^2}{25-49i^2} \\ &= \frac{-11+29i}{74} = -\frac{11}{74} + \frac{29}{74}i\end{aligned}$$

Complex numbers may also be defined as ordered pairs of real numbers. With this approach, the complex number denoted by the expression $a + bi$ is designated by the ordered pair (a,b). Then two complex numbers (a,b) and (c,d) are said to be equal if and only if $a = c$ and $b = d$ and the algebraic operations may be defined as follows:

(1) $(a,b) + (c,d) = (a + c, b + d)$

(2) $(a,b) - (c,d) = (a - c, b - d)$

(3) $(a,b) \cdot (c,d) = (ac - bd, ad + bc)$

(4) $\dfrac{(a,b)}{(c,d)} = \left(\dfrac{ac + bd}{c^2 + d^2}, \dfrac{bc - ad}{c^2 + d^2}\right)$

The fact that this system defined using ordered pairs is completely equivalent to the system defined using expressions of the form $a + bi$ shows that neither the addition sign used in the previous notation nor the use of the imaginary unit i is essential. For example, noting that $(0,1) = i$ and $(-1,0) = -1$, we see that the equation $i^2 = -1$ may be replaced by $(0,1)^2 = (-1,0)$. Also the result shown in Example 2 above may be expressed as

$$(2,3) \cdot (5,-7) = (31,1)$$

1.11 Algebraic Equations

An equation is a statement of equality between two or more algebraic expressions. For example, $7x - 4 = 5x + 6$. The expressions that are equated to each other are referred to as the **sides** of equation and also as the **members** of the equation. The solution set for a given equation is the set of all numbers, each of which, when substituted in the equation, reduces it to a numerical identity, i.e., an equation of the type $a = a$. A number that is a member of the solution set is said to **satisfy** the equation. In the above example, the solution set can be obtained by employing certain elementary operations to produce a series of equivalent equations, i.e., equations having the same solution set as the original equation.

Example 1:

$$7x - 4 = 5x + 6$$

Solution Subtracting $5x - 4$ from both sides of the equation yields $2x = 10$. Dividing both sides of the equation by 2 yields $x = 5$. Substitution of this value of x in both sides of the original equation produces the identity $31 = 31$. Hence the set $\{5\}$ is the solution set for this equation.

Example 2 Consider the equation $2x^2 - 5 = 13$. Again we may obtain the following sequence of equivalent equations:

$$2x^2 - 5 = 13$$

$$2x^2 = 18$$

$$x^2 = 9$$

$$x = \pm 3$$

The solution set is $\{3,-3\}$, either element of which, when substituted in the original equation, produces the identity $13 = 13$.

An equation whose solution set consists of *all* of the numbers that produce a real number when substituted in either side of the equation is called an **identity.** Here are two examples.

Example 3:

$$4(x-3) = 4x - 12$$

Example 4:

$$\left(\frac{1}{x-2}\right)^2 = \frac{1}{x^2 - 4x + 4}$$

The solution set for Example 3 is the set of all real numbers. The solution set for Example 4 is the set of all real numbers except the number 2, which cannot be substituted in either side of Example 4 since division by 0, which is an undefined operation, would be indicated.

Equations like the ones that have been discussed, in which all of the exponents associated with the unknown quantities are positive integers, are called **polynomial equations.** The **degree** of a polynomial equation corresponds to the largest exponent occurring in the equation. For example, the equation $4x^2 - 5x + 17 = 0$ is a second-degree equation, whereas the equation $x^3 - 9x^2 = 10$ is a third-degree equation. Equations of the first degree are usually called **linear equations;** those of the second degree are called **quadratic equations;** those of the third degree are called **cubic equations.**

We have already furnished examples of finding the solution set for linear and quadratic equations. Let us consider next the equation $x^2 + 3x - 10 = 0$, also a quadratic equation. In obtaining our sequence of equivalent equations in this instance, we shall make use of factoring.

Example 5:

$$x^2 + 3x - 10 = 0$$
$$(x+5)(x-2) = 0$$
$$x + 5 = 0 \quad \text{or} \quad x - 2 = 0$$
$$x = -5,\ x = 2$$

The two equations in the third line above are equivalent to the single equation in the second line since the product of two factors is 0 if either of the factors is 0. It can easily be verified that each element of the solution set $\{-5,2\}$, when substituted in the original equation, produces the identity $0 = 0$.

In solving quadratic equations, the factoring procedure used in Example 5 is not always applicable. Since equations of this type occur frequently, it is desirable to have available a general method for finding the solution set. Such a method is provided by a well-known algebraic formula, the **quadratic formula.** It may be stated as follows.

Given the quadratic equation

$$ax^2 + bx + c = 0$$

where a, b, and c are constants and a is not 0, the solution set consists of the elements

$$\frac{-b \pm \sqrt{b^2 - 4ac}}{2a}$$

Although the proof of this formula is not difficult, we shall omit it here.

Example 6 Solve the equation

$$x^2 + 3x - 5 = 0$$

Solution Here $a = 1$, $b = 3$, and $c = -5$. Hence we have

$$x = \frac{-3 \pm \sqrt{9 + 20}}{2}$$

and the solution set is

$$\left\{\frac{-3 + \sqrt{29}}{2}, \frac{-3 - \sqrt{29}}{2}\right\}$$

Example 7:

$$x^2 + 4x + 8 = 0$$

$$x = \frac{-4 \pm \sqrt{16 - 32}}{2}$$

$$= \frac{-4 + \sqrt{-16}}{2}$$

$$= \frac{-4 \pm 4i}{2} = -2 \pm 2i$$

The solution set is $\{-2 - 2i, -2 + 2i\}$. Each of the elements of the solution set is a number of a type referred to in Section 1.10, namely, a complex number.

General methods for the solution of polynomial equations of third and fourth degrees also exist, but we shall not discuss them here. No such general methods exist for equations of degrees higher than fourth.

1.12 Systems of Algebraic Equations

A system of equations is a set of two or more equations, and the solution set for a given system of equations is the intersection of the solution sets of the individual equations.

Example 1 Consider the following system, consisting of two linear equations in the unknown quantities x and y.

$$4x - 3y = 7 \tag{1}$$

$$2x + 5y = 23 \tag{2}$$

Solution Evidently the solution set for Equation (1) consists of an infinite set of ordered pairs since for every value selected for x there is a corresponding value of y such that the ordered pair thus obtained is an element of the solution set. For example, solving Equation (1) for y, we obtain $y = (4x - 7)/3$. If we choose $x = 2$, we obtain $y = \frac{1}{3}$, so that $(2, \frac{1}{3})$ is an element of the solution set. Evidently this process can be continued indefinitely, and the same situation prevails with respect to Equation (2). Our problem is to find the intersection of these two infinite solution sets.

We proceed by multiplying Equation (1) by 5 and Equation (2) by 3. This produces the equivalent system.

$$20x - 15y = 35 \tag{1'}$$

$$6x + 15y = 69 \tag{2'}$$

If we add (1′) and (2′), we obtain

$$26x = 104$$

$$x = 4$$

Substituting this value of x in (1) yields

$$16 - 3y = 7$$

$$3y = 9$$

$$y = 3$$

Hence the ordered pair (4,3) is an element of the solution set of the given system of equations. From the manner in which we have proceeded it is intuitively evident that it is the only element.

Example 2:

$$3x - 5y = 15 \tag{3}$$

$$6x - 10y = 30 \tag{4}$$

Solution Inspection of these equations reveals that (4) may be obtained by multiplying (3) by 2. Hence (3) and (4) are equivalent equations and each has the same infinite solution set.

Example 3:

$$3x - 5y = 15 \tag{5}$$

$$6x - 10y = 45 \tag{6}$$

Solution Inspection of this system reveals that the left-hand side of (6) is twice the left-hand side of (5), whereas the right-hand side of (6) is three times the right-hand side of (5). It is therefore clear that no element of the solution set of (5) can also be an element of the solution set of (6). Hence the intersection of the two solution sets is the empty set; i.e., the system has no solution.

In summary, it can be said that the solution set of a system of two linear equations in two unknowns contains either exactly one element, an infinite number of elements, or no elements.

The procedures shown above can be generalized to include systems of three or more linear equations.

Example 4:

$$2x + 3y - z = -10 \tag{7}$$

$$x - 4y + 2z = 24 \tag{8}$$

$$3x - y - 3z = -6 \tag{9}$$

Solution If Equation (7) is left unchanged and Equation (8) is multiplied by 2, we obtain

$$2x + 3y - z = -10 \tag{7'}$$

$$2x - 8y + 4z = 48 \tag{8'}$$

If Equation (8′) is subtracted from Equation (7), we have

$$11y - 5z = -58 \tag{10}$$

Now if Equation (8) is multiplied by 3 and Equation (9) is left unchanged, we obtain

$$3x - 12y + 6z = 72 \qquad (8'')$$
$$3x - y - 3z = -6 \qquad (9)$$

If Equation (9) is subtracted from Equation (8″), we obtain

$$-11y + 9z = 78 \qquad (11)$$

Equations (10) and (11) now constitute a system of two linear equations in two unknowns. If Equations (10) and (11) are added, we obtain $4z = 20$, or $z = 5$. Substituting this value in Equation (10) yields $11y - 25 = -58$, or $11y = -33$. Hence $y = -3$. If the two values already obtained are substituted in Equation (7), we have $2x - 9 - 5 = -10$. This is equivalent to $2x = 4$, or $x = 2$. Hence the solution set for this system of equations consists of the ordered triple $(2,-3,5)$.

The procedure used in solving the above system can be summarized as follows. First eliminate the same unknown from two pairs of the three equations. This leads to a system of two equations in two unknowns. Then eliminate one of the two remaining unknowns from this system, which leaves a single equation in one unknown. After this equation has been solved, the other two unknows can be found by substituting in one of the equations obtained in the first elimination and finally in one of the equations of the original system.

Clearly, this procedure can be generalized to apply to the solution of systems consisting of any number of equations in the same number of unknowns. However, the procedure can be further systematized by using an algebraic entity called a matrix. Another method, called Cramer's rule, makes use of a quantity associated with a matrix called the determinant of the matrix. Both of these will be discussed in Chapter 8.

1.13 Algebraic Inequalities

We now proceed to the discussion of algebraic inequalities, a subject that has assumed increasing importance in recent years. An algebraic inequality is a statement involving two or more algebraic expressions and the signs of inequality, for example, $2x - 3 < 4x + 5$. The expressions $2x - 3$ and $4x + 5$ are called the **members** of the inequality. The solution set of an algebraic inequality is the set of numbers, each of which when substituted in the inequality produces an absolute inequality, i.e., one that is always true, such as $7 > 5$.

In obtaining the solution set for a given inequality, we proceed by obtaining a sequence of equivalent inequalities. In so doing, we make use of the following operations:

(1) Add the same number to or subtract the same number from each member of the inequality.

(2) Multiply or divide each member of the inequality by the same nonzero number, except that if the number is negative, the direction or sense of the inequalities must be reversed.

The validity of operation (1) is rather obvious. For example, adding 5 to both members of the inequality $6 > 4$ produces the inequality $11 > 9$. To illustrate operation (2) note that multiplying both members of the inequality $5 > 2$ by 3 produces the inequality $15 > 6$. However, if both members of the same inequality are multiplied by -3, the direction of the inequality sign must be reversed, to produce the inequality $-15 < -6$.

Example 1 Find the solution set for the inequality

$$3x - 4 > 11$$

Solution Adding 4 to both members yields

$$3x > 15$$

Dividing both members by 3, we have

$$x > 5$$

By use of interval notation, the solution set is found to be $(5,\infty)$.

Example 2:

$$-2x - 4 \leq 12$$
$$-2x \leq 16$$
$$x \geq 8$$

The solution set is $[8,\infty)$.

Example 3:

$$|x - 5| < 11$$

Solution Consider first the equation $|x - 5| = 11$. From the definition of absolute value (Section 1.2) it follows that this equation is equivalent to the set of two equations $x - 5 = 11$ and $x - 5 = -11$, so that the solution set is $\{-6,16\}$.

These considerations lead to the conclusion that the inequality $|x - 5| < 11$ is equivalent to the inequality

$$-11 < x - 5 < 11$$

Adding 5 to each member of this inequality, we obtain

$$-6 < x < 16$$

so that the solution set is the open interval $(-6,16)$.

Example 4:

$$|3x + 5| \geq 20$$

Solution From the previous example, it should be clear that this inequality is equivalent to the set of two inequalities

$$3x + 5 \leq -20 \qquad 3x + 5 \geq 20$$

These inequalities are equivalent, respectively, to the inequalities

$$3x \leq -25 \qquad 3x \geq 15$$

$$x \leq \frac{-25}{3} \qquad x \geq 5$$

Hence, by use of interval notation, the solution set is found to be $(-\infty, \frac{-25}{3}] \cup [5,\infty)$.

Example 5:

$$x^2 - 16 < 0$$
$$x^2 < 16$$

Solution The last inequality is evidently equivalent to $|x| < 4$, or $-4 < x < 4$, so that the solution set is the interval $(-4,4)$.

Example 6:

$$x^2 - 3x - 10 < 0$$

Solution The left-hand side can be factored, to give

$$(x + 2)(x - 5) < 0$$

Now the product of two factors is negative if and only if they are opposite in sign. Hence, the solution set for this last inequality consists of all values of x for which either

$$(x + 2) > 0 \text{ and } (x - 5) < 0$$

or $$(x+2)<0 \text{ and } (x-5)>0$$

In the first case, we have $x > -2$ and $x < 5$, so that the interval $(-2,5)$ is included in the solution set. In the second case, we have $x < -2$ and $x > 5$. Since this is clearly impossible, the interval $(-2,5)$ constitutes the entire solution set.

The solution of systems of inequalities in two or more variables is a subject of considerable practical importance. However, because a graphical approach is necessary in this case, this topic will be discussed in connection with graphs in Chapter 2.

1.14 Elementary Counting Procedures

The most elementary counting procedure is direct enumeration, i.e., counting individually each object in a group of objects, whose number of members is desired. However, if the number of objects is very large, this procedure is laborious and time-consuming. In many cases, more sophisticated and vastly more efficient counting procedures are available.

The simplest of these procedures is exemplified in the following simple problem. If a man has 4 shirts and 5 ties, how many different combinations of a shirt and tie can he wear? Evidently since he can wear 1 of his 5 ties with each of his 4 shirts, he can wear $5 \cdot 4 = 20$ different combinations of these articles. The principle that has been employed is well-known and frequently used. For example, to calculate the area of a rectangle, we multiply the length by the width in order to find the number of square units of area contained in the rectangle. Furthermore, this multiplication procedure often involves products of more than two factors. For example, if the man referred to above has 3 suits, then we find that he may wear $5 \cdot 4 \cdot 3 = 60$ different combinations of a shirt, a tie, and a suit.

The counting principle involved in these examples may be stated formally as follows:

If an action can be performed in n_1 ways, a second action in n_2 ways, a third action in n_3 ways, etc., then the total number of ways in which k actions can be performed in a specified order is $n_1 \cdot n_2 \cdot \cdots \cdot n_k$.

Note that although a particular order is specified in the application of the above principle, in many cases the order is immaterial. For example, the order in which a man selects a shirt and a tie does not affect the number of different shirt and tie combinations which he can wear.

1.15 Permutations

We introduce the idea of permutations with an example. If there are 6 teams in a baseball league, in how many different orders may the teams finish at the end of the season? We may proceed with this calculation as follows: The first place in the order of finish may be filled by any of the 6 teams. For each of these 6 possible first-place finishes there are 5 possible second-place finishes. Hence, applying the counting principle stated in the previous section, we find that there are $6 \cdot 5 = 30$ possible first- and second-place finishes. Continuing in the same way for the other places, we see that there are $6 \cdot 5 \cdot 4 \cdot 3 \cdot 2 \cdot 1 = 720$ possible orders in which the 6 teams may finish.

The product $6 \cdot 5 \cdot 4 \cdot 3 \cdot 2 \cdot 1$ is often abbreviated by the use of the symbol 6! (read as *six factorial*). In general for any positive interger n we define **n factorial** as

$$n! = n(n-1)(n-2) \cdot \cdot \cdot 1^*$$

In addition, it is convenient to extend this definition to include $n = 0$ by defining $0! = 1$.

In the previous example, we calculated the number of different orders in which 6 teams can be arranged. Each such arrangement is called a **permutation.** Clearly the line of reasoning we have employed in this example can be generalized, and we can state the following.

Theorem I. The number of permutations of n distinct objects is $n!$

Let us now consider a slight variation on the previous example. If there are 6 teams in a baseball league, in how many different orders can the teams finish first, second, third, and fourth? Evidently, using the same reasoning as before but taking into account the fact that there are now only 4 positions to fill, we obtain $6 \cdot 5 \cdot 4 \cdot 3 = 360$ as the answer to our problem. Note that in this case we have found all possible arrangements of 6 different objects using only 4 of them in each arrangement. Furthermore, we can think

* Three dots are often used as a mathematical symbol for *etc.*

of the product we have used to calculate our answer as being of the form

$$(6-0)\cdot(6-1)\cdot(6-2)\cdot(6-3)$$

Once again we can generalize the procedure used in this last example. If we have available n different objects, then the number of arrangements of these objects using r of them in each arrangement is

$$n(n-1)(n-2)\cdots[n-(r-1)] = n(n-1)(n-2)\cdots(n-r+1)$$

However, this last product may be written more compactly using factorial notation:

$$\begin{aligned} &n(n-1)(n-2)\cdots(n-r+1) \\ &= \frac{n(n-1)(n-2)\cdots(n-r+1)(n-r)!}{(n-r)!} \\ &= \frac{n(n-1)(n-2)\cdots(n-r+1)(n-r)(n-r-1)(n-r-2)\cdots 1}{(n-r)!} \\ &= \frac{n!}{(n-r)!} \end{aligned}$$

Hence we have the following.

Theorem II. The number of permutations of n different objects taken r at a time, denoted by the symbol ${}_nP_r$, is $n!/(n-r)!$.

Note that $r \leq n$ and if $r = n$, the expression $n!/(n-r)!$ becomes $n!/0! = n!$. Hence, Theorem I is a special case of Theorem II.

Example 1 If any arrangement of a set of letters is called a word, how many 4-letter words can be formed from the 26 letters of the alphabet?
Solution We wish to find the number of permutations of 26 things taken 4 at a time. Using Theorem II, with $n = 26$ and $r = 4$, we calculate ${}_{26}P_4 = 26!/22! = 26 \cdot 25 \cdot 24 \cdot 23 = 358{,}800$.

We discuss next the necessary modification of the formula given in Theorem I for the number of permutations of n distinct objects if some of the objects are in some respect identical. For example, suppose we wish to determine the number of possible arrangements of 7 colored balls that are to be placed in a row if 2 of the balls are of the same color and the remaining balls are of different colors.

We may reason as follows. If the 2 balls of the same color were somehow distinguished from each other, then by Theorem I the number of possible arrangements would be 7!. However, for each of these 7! arrangements there would be another arrangement that would be identical except that the 2 balls of the same color would be interchanged. Hence, the number of possible arrangements if the 2 balls of the same color are not distinguished from each other is 7!/2!. By the same line of reasoning, if 3 of the 7 balls were of the same color, the number of different arrangements would be 7!/3!. Generalizing still further to cases in which there is more than one set of objects that are not distinct from each other, the following theorem may be proved.

Theorem III. The number of permutations of n objects, of which n_1 are alike of one kind, n_2 are alike of another kind, . . . , and n_k are alike of the kth kind, is

$$\frac{n!}{n_1!n_2! \cdot \cdot \cdot n_k!}$$

Example 2 Find the number of permutations of the letters of the word "Mississippi."

Solution The 11 letters include 4 i's, 4 s's, and 2 p's. Hence the number of permutations is

$$\frac{11!}{4!4!2!} = 34{,}650$$

1.16 Combinations

In many cases we wish to determine the number of different selections of r objects that can be made from a group of n objects without regard to the order in which they are arranged. For example, consider the problem of determining the number of committees of 3 persons that can be selected from a group of 7 persons. Clearly, changing the order in which names of the persons comprising a given committee are arranged does not change the committee. Order is of no importance.

A selection of the type referred to above is called a combination. In order to lead up to a formula for determining the number of combinations of n things taken r at a time, we shall consider a simple example. Consider the set of 3 objects a, b, and c. There are 6 permutations of these objects taken 2 at a time:

$$ab, ba, ac, ca, bc, cb$$

However, in view of the definition of combinations given above, there are evidently only 3 combinations of these objects taken 2 at a time:

$$ab,\ ac,\ bc$$

It is readily seen that the number of permutations, in this case 6, can be obtained by multiplying the number of combinations, 3, by the number of orders, 2, in which each selection of 2 can be arranged.

In general, when r objects are selected from n, there are $r!$ orders in which each selection can be arranged. Hence, the number of permutations can be obtained by multiplying the number of combinations by $r!$ If the number of combinations is denoted by

$$\binom{n}{r}$$

we have

$${}_nP_r = \binom{n}{r} \cdot r!$$

Hence $$\binom{n}{r} = \frac{{}_nP_r}{r!} = \frac{\dfrac{n!}{n-r!}}{r!} = \frac{n!}{r!(n-r)!}$$

Theorem IV. The number of combinations of n different objects taken r at a time is

$$\frac{n!}{r!(n-r)!}$$

To apply Theorem IV, consider the previously stated problem of finding the number of committees of 3 that can be selected from a group of 7 persons. Evidently this is

$$\binom{7}{3} = \frac{7!}{3!4!} = \frac{7 \cdot 6 \cdot 5}{3 \cdot 2 \cdot 1} = 35$$

1.17 Subscript and Summation Notation

In some cases the same letter is used to designate several different numbers, the values being distinguished by means of subscripts, i.e., numbers written at the lower right of the letter. For example, five different numbers might be designated by x_1, x_2, x_3, x_4, and x_5, which would be read as x sub-one, x sub-two, etc.

Furthermore, the sum of these five numbers, i.e., $x_1 + x_2 + x_3 + x_4 + x_5$, is frequently designated as follows:

$$\sum_{i=1}^{5} x_i$$

This expression, which employs the Greek capital letter sigma, is described as the summation of x sub-i from 1 to 5. The sum $x_1 + x_2 + \cdots + x_n$ is designated by

$$\sum_{i=1}^{n} x_i$$

Here are some additional examples of the use of the summation symbol.

$$\sum_{i=1}^{n} x_i^2 = x_1^2 + x_2^2 + \cdots + x_n^2$$

$$\sum_{i=1}^{n} x_i y_i = x_1 y_1 + x_2 y_2 + \cdots + x_n y_n$$

$$\sum (x_i - y_i) = (x_1 - y_1) + (x_2 - y_2) + \cdots + (x_n - y_n)$$

The infinite sum $x_1 + x_2 + \cdots$ is designated by

$$\sum_{i=1}^{\infty} x_i$$

In many cases, in which the range of the summation is clear from the context, it is simply omitted. For example, the symbol Σx stands for the sum of all possible values of x under consideration.

In the algebraic manipulation of summation symbols, three important facts about summations are often used. They are stated below without proof. In (2) and (3), k is a constant, and in (3), the summation is assumed to be from 1 to n.

(1) $\Sigma(x + y) = \Sigma x + \Sigma y$

(2) $\Sigma kx = k\Sigma x$

(3) $\Sigma k = nk$

Problems—Chapter 1

Sets

1. Given

$$U = \{1,3,5,7,9,11,13,15\}$$
$$A = \{1,5,9,11\},\ B = \{5,9,13,15\}$$

find the following sets:

(a) $A \cup B$
(b) $A \cap B$
(c) $A - B$
(d) A'
(e) $A' \cup B'$
(f) $(A \cap B)'$
(g) $A' \cap B'$
(h) $(A \cup B)'$
(i) $A \times B$
(j) $B \times A$

Note that parts (e), (f), (g), and (h) illustrate the deMorgan laws.

2. In the Venn diagram below, the number of elements in each of a number of subsets is indicated. None of the numbers is included in any other number. For example, the total number of elements in the set A is $7 + 5 + 4 + 2 = 18$.

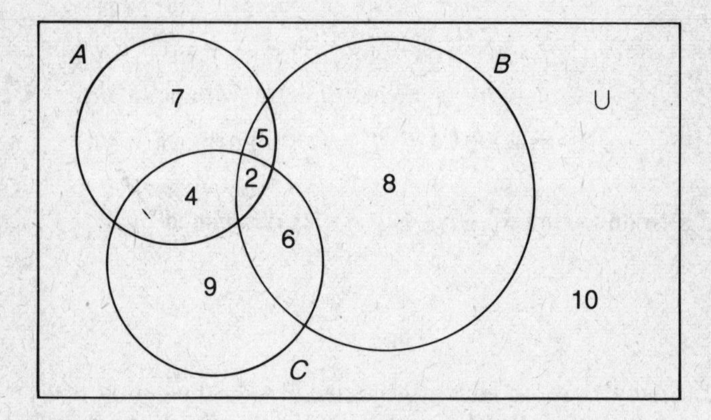

Find the number elements in each of the following sets:

(a) $A \cup B$
(b) $A \cap C$
(c) A'
(d) $A \cup B'$
(e) $B \cap C'$
(f) $A' \cup B'$
(g) $A' \cap B'$
(h) $A \cap B \cap C'$
(i) $A \cap B' \cap C'$

Exponents and Radicals

3. Carry out the indicated operations and express without negative exponents:

(a) $(x^2y^{-3})^{-4}$

(b) $\dfrac{x^5y^3z^0}{x^7yz^{-1}}$

(c) $[(x^2y^{-2}) \cdot (x^{1/3}y^4)]^{-2}$

4. Evaluate:
   (a) 5^{-3}
   (b) $(\frac{2}{3})^{-2}$
   (c) $(125)^{5/3}$
   (d) $(16)^{-7/4}$
5. Perform the indicated operations and simplify:
   (a) $\sqrt[3]{56} + 4\sqrt[3]{7}$
   (b) $\sqrt[4]{32x^6y^8}$ $\quad x > 0, y > 0$
   (c) $\sqrt{50x^2}$ $\quad x < 0$
   (d) $\sqrt[3]{x} \cdot \sqrt[5]{x}$
   (e) $\dfrac{\sqrt{x}}{\sqrt[4]{x}}$ $\quad x \neq 0$

Factoring

6. Factor the following expressions:
   (a) $16x^4 - 81y^4$
   (b) $(x^6 + 64y^6)$
   (c) $(x^{15} - y^{12})$
   (d) $cx - dy - cy + dx$
   (e) $ax^3 - by^3 - ay^3 + bx^3$

Variation

7. Find the law of variation in each of the following cases:
   (a) A quantity y varies directly with a quantity x; when $x = 2$, $y = -6$.
   (b) A quantity y varies inversely with a quantity x; when $x = \frac{1}{2}$, $y = \frac{3}{8}$.
   (c) A quantity q varies directly with the product of the quantities x and y and inversely with the quantity z; when $q = 12$, $x = 2$, $y = 3$, and $z = 4$.

Complex Numbers

8. Carry out the indicated operations and express in the form $a + bi$:
   (a) $(3 + 4i) + (2 - 5i)$
   (b) $(3 + 4i) \cdot (2 - 5i)$
   (c) $(3 + 4i) \div (2 - 5i)$
   (d) $\dfrac{1}{7-i}$
   (e) $(2 + i)^3$

Solution of Equations

9. Solve each of the following equations:
   (a) $2x - 5 = 7x - 15$
   (b) $3x^2 - 7 = 41$
   (c) $x^2 - 4x - 5 = 0$
   (d) $x^3 - 3x^2 - 28x = 0$
   (e) $x^2 - 10x + 25 = 0$
   (f) $x^2 + 4x + 2 = 0$
   (g) $x^2 + 6x + 10 = 0$

10. Solve each of the following systems of equations:
    (a) $2x + 3y = 14$
    $x - 4y = -26$
    (b) $3x + 4y = 7$
    $6x + 8y + 15$
    (c) $2x + 3y - z = 19$
    $3x - 4y + z = -2$
    $x + y + 2z = 3$

Algebraic Inequalities

11. Solve each of the following inequalities:
    (a) $4x - 5 < 19$
    (b) $2x + 7 \geq -5$
    (c) $|x + 3| < 7$
    (d) $|5x - 2| \geq 13$
    (e) $x^2 - 7x + 10 < 0$
    (f) $x^2 - 6x + 9 \geq 0$

Permutations and Combinations

12. How many different arrangements of 4 balls each may be made using 10 balls of different colors?
13. Find the number of permutations of the letters of the word "Alabama."
14. In how many ways may a committee of 4 be selected from a group of 9 persons?
15. A repair service has 10 plumbers.
    (a) If the plumbers work in teams of 3, how many different repair teams can be formed?
    (b) How many of these teams will include one specified plumber?

Summation Notation

16. Given

$$x_1 = 2,\ x_2 = 3,\ x_3 = 5$$
$$y_1 = 1,\ y_2 = 4,\ y_3 = 6$$

Find:
    (a) Σx
    (b) Σx^2
    (c) $\Sigma x \Sigma y$
    (d) Σxy
    (e) $\Sigma x^2 y$

2

FUNCTIONS AND GRAPHS

2.1 Functions in General

The notion of a function is perhaps the most important idea in all of mathematics. Basically a **function** is simply a relationship between two sets. One set is called the **domain** of the function, the other set is called its **range.** The essence of a function is a rule that associates with each element of the domain one and only one element of the range. Each of the elements of the range corresponds to at least one element of the domain.

A function may thus be specified simply by a set of ordered pairs in which the first element of each pair is an element of the domain, the second is an element of the range, and for which no two pairs have the same first element. For example, one function might be defined as follows:

$$\{(2,4), (5,9), (3,7), (1,10)\}$$

Note that the following set of ordered pairs is not a function:

$$\{(2,4), (2,9), (3,7), (1,10)\}$$

since corresponding to the element 2 of the domain there are two elements of the range.

A correspondence in which more than one element of the range may be associated with a specified element of the domain is called a **relation.** Hence, a function is a special case of a relation. The second set of ordered pairs shown above evidently defines a relation.

Most of the useful functions in mathematics are defined by means of equations, for example, $y = x^2$. Note that this equation may be used to generate an infinite set of ordered pairs of the form (x,y), for example, $\{(0,0)\ (1,1)\ (2,4), (-3,9), \ldots\}$, which satisfies the definition of a function as stated above. Note that the domain of this function may be taken as the set of all real numbers and that the range then consists of the set of all non-negative real numbers.

It should also be noted that the domain may be restricted when a function is defined. In the above example, the domain could be arbitrarily chosen to be the set $\{x|-3 \leq x \leq 3\}$; the range of the function would then consist of the set $\{y|0 \leq y \leq 9\}$.

Unless the domain of a function is explicitly specified, it is usually assumed that it consists of the set of all real numbers whose corresponding numbers in the range are also real. Let us consider some additional examples. Both set and interval notation are shown.

	DOMAIN	RANGE
Example 1:		
$y=\sqrt{x-5}$	$\{x\|x \geq 5\}$ $[5,\infty)$	$\{y\|y \geq 0\}$ $[0,\infty)$
Example 2:		
$y=-\sqrt{x^2-9}$	$\{x\| \|x\| \geq 3]$ $(-\infty,-3] \cup [3,\infty)$	$\{y\|y \leq 0\}$ $[0,\infty)$
Example 3:		
$y=\dfrac{1}{x-4}$	$\{x\|x \neq 4\}$ $(-\infty,4) \cup (4,\infty)$	$\{y\|y \neq 0\}$ $(-\infty,0) \cup (0,\infty)$
Example 4:		
$y=\dfrac{1}{\sqrt{16-x^2}}$	$\{x\|-4 < x < 4\}$ $(-4,4)$	$\{y\|\frac{1}{4} \leq y < \infty\}$ $[\frac{1}{4}, \infty)$

These examples should be carefully studied. Note that in Example 1 a value of x less than 5 would produce an imaginary y. Furthermore, y can assume no non-negative values in view of the definition of the symbol $\sqrt{}$. In Example 2 any value of x in the open interval $-3 < x < 3$ would produce imaginary values of y. In Example 3 the value $x = 4$ would produce an undefined y. It is also clear that y cannot be 0 in view of the fact that for a fraction to be 0 it is necessary for the numerator to be 0. In Example 4 it is clear that the smallest value of y, namely, $\frac{1}{4}$, occurs when $x = 0$ and that y increases without limit as x approaches either 4 or -4.

To return to a previous example, an alternate way of representing the function $y = x^2$ is $f(x) = x^2$.* This is an example of the very useful notation called **functional notation.** By use of this symbol the element of the range corresponding to the element 2 of the domain can be represented by the notation $f(2)$. Here $f(2) = 4$. Also $f(0) = 0$ and $f(-6) = 36$. The elements of the range are often called simply the values of the function. If two or more functions are under discussion at the same time, then they may be represented by different letters. For example, we might specify $g(x) = x^3 - 4$ and $h(x) = \sqrt{1/x}$. In connection with these examples, we would have $g(1) = -3$ and $h(9) = \frac{1}{3}$.

Furthermore, functional notation is sometimes used in the construction of what is called a **composite function.** For example, the composite function $f[g(x)]$ is obtained by substituting $g(x)$ for x in the function $f(x)$. Thus, if $f(x) = x^2$ and $g(x) = x^3 - 4$, then $f[g(x)] = (x^3 - 4)^2 = x^6 - 8x^3 + 16$. This function is called the **composition** of g by f. On the other hand, the composition of f by g is given by $g[f(x)] = (x^2)^3 - 4 = x^6 - 4$.

The idea of composition of functions can be extended to cases involving more than two functions. For example, if $f(x) = 1/x$, $g(x) = \sqrt{x}$, and $h(x) = x^3$, then

$$g[h(x)] = \sqrt{x^3} = x^{3/2}$$

and

$$f\{g[h(x)]\} = \frac{1}{x^{3/2}}$$

In some cases more than one equation is needed to specify a function completely. Consider the following example. A taxicab company charges 50 cents/mi for a trip of 5 mi or less and 25 cents/mi for each mile in excess of 5. Express the cost C, in cents, of a trip in terms of the number of miles x. Clearly for a trip of 5 mi or less the cost will be $50x$. For a trip of more than 5 mi the cost will be the cost of the first 5 mi, or 250, plus the mileage in excess of 5 multiplied by 25. Hence the cost will be given by $250 + 25(x - 5) = 25x + 125$, and the complete specification of the function will be given by

$$C = \begin{cases} 50x & x \leq 5 \\ 25x + 125 & x > 5 \end{cases}$$

* Some texts distinguish between the symbols f and $f(x)$ by denoting the function, i.e., the set of ordered pairs, by f and a value of the function by $f(x)$. However, we shall use both the symbols f and $f(x)$ to denote the function.

2.2 The Graph of a Function

The **graph of a function** is a pictorial representation of a function that is obtained by using the two-dimensional rectangular coordinate system previously introduced. More precisely, the graph of a function is the set of all points in the plane, each of whose coordinates constitute one of the ordered pairs that define the function.

The graph of a function may be obtained by constructing a table of corresponding elements of the domain (x) and the range (y) of the function. Each of these pairs of values determines a point in the plane. Each of these points may then be *plotted* by placing a dot in the plane at the appropriate location. Since the number of such points is, in most cases, infinite, it is impossible to plot all of them. We must therefore be content with plotting a representative sample of them and then joining them by means of a smooth curve.

Here are some examples, including some of the functions that have already been discussed.

Example 1:

$y = x^2$

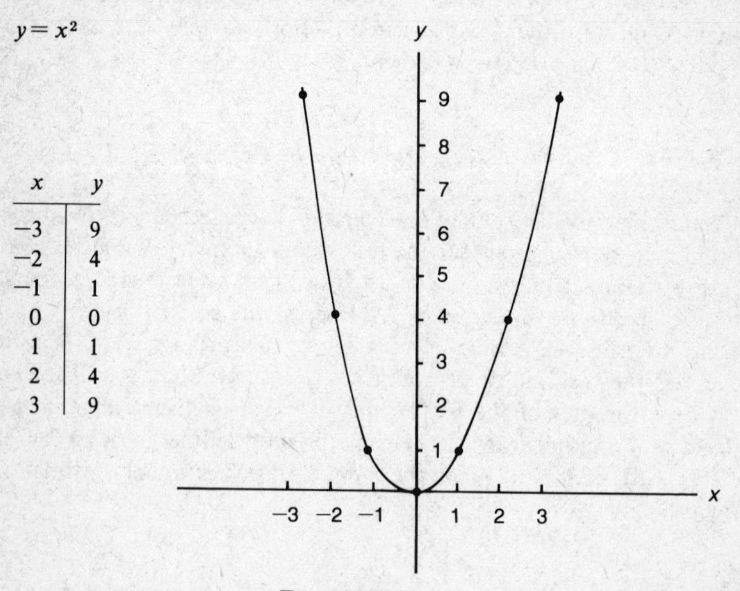

x	y
−3	9
−2	4
−1	1
0	0
1	1
2	4
3	9

FIGURE 2.1(1)

Example 2:

$y = \sqrt{x-5}$

x	y
5	0
6	1
9	2
14	3

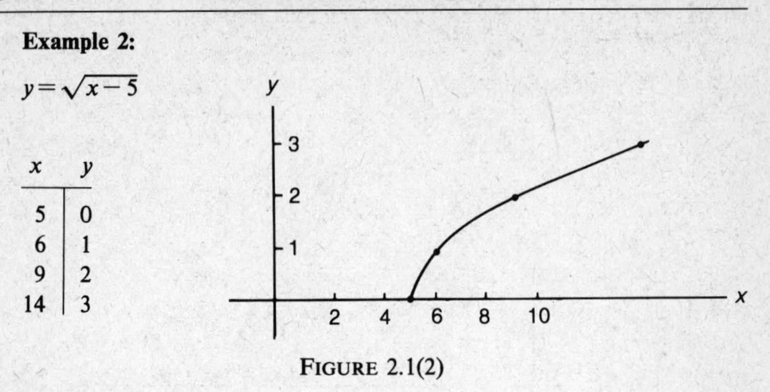

FIGURE 2.1(2)

Example 3:

The cost function $C = \begin{cases} 50x & x \leq 5 \\ 25x + 125 & x > 5 \end{cases}$

x	c
0	0
1	50
2	100
5	250
7	300
10	375

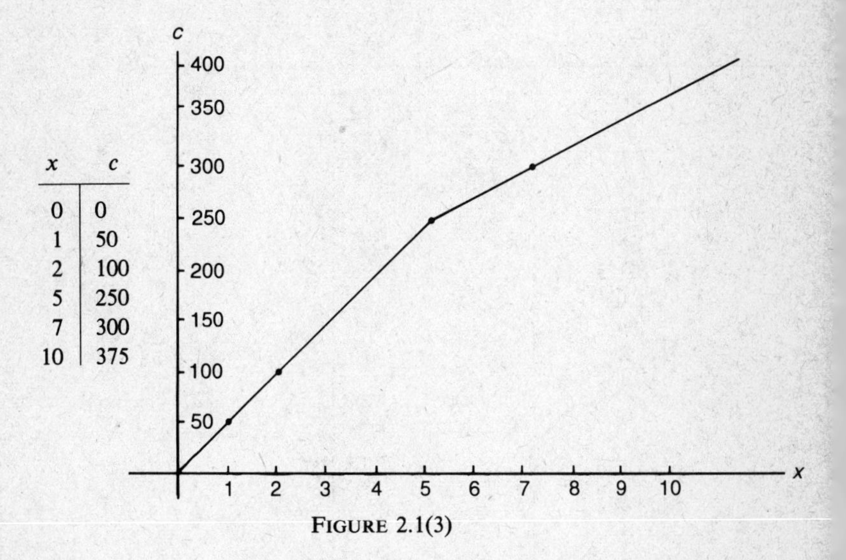

FIGURE 2.1(3)

Example 4:

$y=|x-4|$

x	y
0	4
2	2
4	0
6	2
8	4

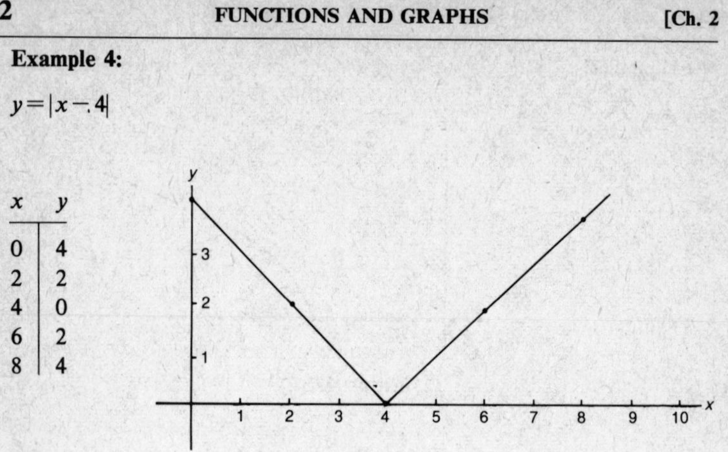

FIGURE 2.1(4)

Example 5:

$$y=\frac{1}{x-4}$$

x	y
0	−1/4
1	−1/3
2	−1/2
3	−1
3.5	−2
3.75	−4
4.25	4
4.5	2
5	1
6	1/2
7	1/3

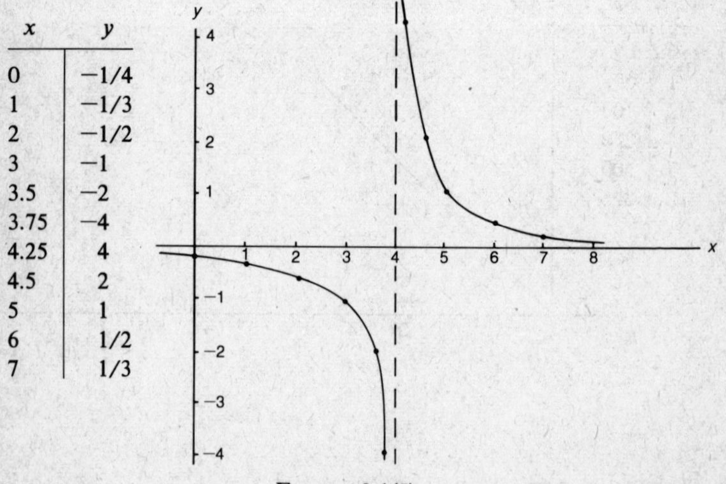

FIGURE 2.1(5)

Note that in the graph of Example 5 above, the curve never touches the x axis since y can never assume the value 0. Note also that the curve is divided into two distinct branches and that each branch approaches the broken line that is drawn parallel to the y axis and that passes through the point (4,0). However, neither branch ever touches this line since x can never assume the value 4. The x axis and the broken line are called **asymptotes** of the curve.

It should also be pointed out that all of the functions discussed thus far are called algebraic. Later on we shall discuss other types of functions, of great importance in mathematics, that are usually called **transcendental.** These include the trigonometric, inverse trigonometric, logarithmic, and exponential functions.

2.3 The Limit of a Function

To understand the idea of a **limit,** consider the function $f(x) = \sqrt{|x-5|}$. It seems evident that as the values of x approach 5, the values of the function approach 0. In fact, it is clear that the value of the function can be brought as close to 0 as desired by choosing a value of x sufficiently close to 5. In this case the function is said to have the limit 0 as x approaches 5. The concept of a limit will now be stated more precisely.

A function $f(x)$ is said to have the limit (lim) L as x approaches p if for every positive number q, however small, there exists a positive number r such that $|f(x) - L| < q$ whenever $0 < |x - p| < r$. Under these conditions we write

$$\lim_{x \to p} f(x) = L$$

The value of the number r in general depends on the value of q. For example, in the case of function $f(x) = \sqrt{|x-5|}$, suppose we choose $q = 0.1$; i.e., we require that the difference between $\sqrt{|x-5|}$ and 0 be less than 0.1. This requirement is met whenever $0 < |x - 5| < 0.01$. Hence, in this case, $r = 0.01$ is a suitable choice. Since a value of r could be found for any value of q, however small, the limit definition is satisfied and $\lim_{x \to 5} \sqrt{|x-5|} = 0$.

It should be pointed out that the concept of a function approaching a limit as x approaches p does not in any way involve the particular value of the function corresponding to $x = p$. Hence, the fact that the value of the function $f(x) = \sqrt{|x-5|}$ is 0 when $x = 5$ is not pertinent to the discussion concerning the limit of the function.

It is for this reason that in the above definition we say "whenever $0 < |x - p| < r$" instead of simply "$|x - p| < r$."

This idea may be clarified by another example. Consider the function

$$f(x) = \frac{x^2 - 9}{x - 3} \qquad x \neq 3$$

This function has no value at $x = 3$, since in this case both the numerator and the denominator assume the value 0, and division by 0 is undefined. However, since for all $x \neq 3$,

$$\frac{x^2 - 9}{x - 3} = \frac{(x + 3)(x - 3)}{x - 3} = x + 3$$

we see that if $x \neq 3$, the function

$$f(x) = \frac{x^2 - 9}{x - 3}$$

is identical to the function $g(x) = x + 3$; hence it follows that

$$\lim_{x \to 3} f(x) = 6$$

The graphs of $f(x)$ and $g(x)$ are shown in Figure 2.2.

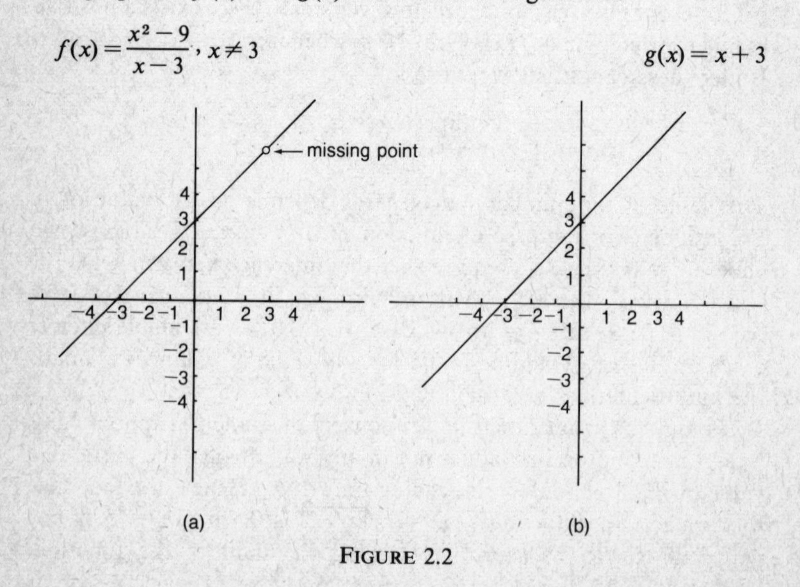

FIGURE 2.2

Once the concept of the limit of a function has been defined, we are in a position to investigate certain properties of limits. Some of the more important of these properties are listed below without proof. In each case "lim" means lim $x \to p$.

(1) $\lim kf(x) = k \lim f(x)$, where k is a constant

(2) $\lim\, [f(x) + g(x)] = \lim f(x) + \lim g(x)$

(3) $\lim\, [f(x) \cdot g(x)] = \lim f(x) \cdot \lim g(x)$

(4) $\lim \dfrac{f(x)}{g(x)} = \dfrac{\lim f(x)}{\lim g(x)}$ if $\lim g(x) \neq 0$

Thus far we have been concerned only with the limit of a function $f(x)$ as x approaches a finite number. However, the limit concept is readily adapted to a situation in which x increases without limit. The limit, if it exists, is indicated in this case by the symbol $\lim_{x \to \infty} f(x)$. It should be emphasized that the symbol ∞ does not stand for a number. The expression $x \to \infty$ is simply a mathematical shorthand for the statement, x is increasing in such a way that it will eventually exceed any prescribed number, however large.

In this case our previous definition must be modified as follows: A function $f(x)$ is said to approach a limit L as x increases without limit if for every positive number q, however small, there exists a positive number r such that $|f(x) - L| < q$ whenever $x > r$.

For an illustration of this extension of the concept of limit consider the function

$$f(x) = 5 + \frac{1}{x}$$

As x increases, the fraction $1/x$ decreases; it is clear that its value can be made as close to 0 as desired by choosing x sufficiently large. Hence $\lim_{x \to \infty} (5 + 1/x) = 5$. In terms of the definition given above, suppose we choose $q = 0.01$, i.e., we wish to make $|(5 + 1/x) - 5| < 0.01$. Clearly this can be done by choosing r as 100.

For a second example of this type, consider the function

$$f(x) = \frac{4x - 3}{2x + 5}$$

Note that as x increases without limit, so do both the numerator and the denominator of $f(x)$. Consequently it is not immediately clear whether the quotient is approaching a limit. However, if we divide both the numerator and the denominator of $f(x)$ by x, we obtain a function

$$g(x)=\frac{4-3/x}{2+5/x}$$

that is identical to $f(x)$ for all x except $x=0$. It is now apparent that $\lim_{x\to\infty}(4-3/x)=4$ and $\lim_{x\to\infty}(2+5/x)=2$. Hence, in view of property (4) above, we have

$$\lim_{x\to\infty} g(x)=\lim_{x\to\infty} f(x)=\frac{4}{2}=2$$

2.4 Continuity

In the preceding discussion of the concept of limit, it was emphasized that the question of whether a function $f(x)$ approaches a limit as x approaches a number p does not in any way involve the value of the function when $x=p$. When the limit of the function does indeed exist, there are three possibilities with regard to the value of the function at $x=p$: The function may have no value at all at $x=p$; it may have a value different from the limit of the function; or it may have a value equal to the limit. The last case provides the basis for the definition of an important property of certain functions called **continuity.** A function $f(x)$ is said to be continuous at a point $x=p$ if

$$\lim_{x\to p} f(x)=f(p)$$

Note that for continuity at a point $x=p$ three things are required: The limit of the function as x approaches p must exist; the function must have a value at $x=p$; and these two quantities must be equal.

A function $f(x)$ that is continuous at all points of an interval, for example, the interval $[0,1]$, is said to be continuous in the interval. A function that is continuous for all values of x is referred to simply as a **continuous function.** An important class of continuous functions is the set of polynomial functions. These are functions of the form

$$f(x)=a_0x^n+a_1x^{n-1}+\cdots+a_nx+a_{n+1}$$

where the a's are constants and all of the exponents are positive integers. For example, the function $f(x)=8x^3+3x-7$ is a polynomial function and is therefore continuous. On the other hand consider the function

$$f(x) = \begin{cases} \dfrac{x+4}{x-5} & x \neq 5 \\ 0 & x = 5 \end{cases}$$

This function has a value at $x = 5$, but $\lim_{x\to 5} f(x)$ does not exist, so it is discontinuous at that point, although it can be shown to be continuous at all other points. The function

$$f(x) = \frac{x^2 - 9}{x - 3}$$

(Figure 2.2) is discontinuous at $x = 3$. The function

$$f(x) = \frac{1}{x - 4}$$

(Figure 2.1(5)) is discontinuous at $x = 4$.

2.5 The Inverse of a Function

It will be recalled that it is basic to the definition of a function that to each element of the domain there corresponds one and only one element of the range. For some functions the reverse is also true, so that no two elements of the domain correspond to the same element of the range. Such functions are called **one-to-one functions.** For example, the function

$$f = \{(1,4),\ (2,7),\ (5,8)\}$$

is a one-to-one function, whereas the function

$$g = \{(1,4),\ (2,4),\ (5,8)\}$$

is not one-to-one.

With each one-to-one function, there is associated a function called the **inverse** of the given function, which is obtained by reversing each of the ordered pairs that constitute the function. The inverse of a function f is denoted by the symbol f^{-1}. Thus for the function f above, we have

$$f^{-1} = \{(4,1),\ (7,2),\ (8,5)\}$$

Note, however, that if we reverse the ordered pairs in the function g, we obtain the set

$$\{(4,1),\ (4,2),\ (8,5)\}$$

which is not a function. This example makes it clear that only one-to-one functions have inverses.

Consider now the function $f(x) = x^3 - 5$. This function is one-to-one and therefore has an inverse. It may be obtained as follows. For convenience we write the function in the form $y = x^3 - 5$. In order to reverse the ordered pairs that constitute this function, we simply interchange x and y, to obtain $x = y^3 - 5$. Now, to obtain the inverse function, we solve this equation for y as follows:

$$x = y^3 - 5$$

$$y^3 = x + 5$$

$$y = \sqrt[3]{x+5}$$

It can easily be verified that by means of the algebraic manipulation carried out above, we have in fact reversed each of the ordered pairs that constitute the original function. For example, the function $f(x) = x^3 - 5$ contains the ordered pair (2,3), whereas the substitution of 3 for x in the function $f^{-1}(x) = \sqrt[3]{x+5}$ generates the ordered pair (3,2).

2.6 Linear Functions

A simple but important type of function is the type known as **linear.** A linear function is one that can be expressed in the form

$$f(x) = mx + b$$

where m and b are constants. A function of this type is called linear because its graph is always a straight line, a fact that will presently be demonstrated.

Note that any equation of the form $Ax + By + C = 0$, $B \neq 0$, implicitly defines a linear function since if the equation is solved for y, it becomes $y = (-A/B)x - C/B$, which is identical in form to the equation above, since $f(x)$ has been replaced by y, m by $-A/B$, and b by $-C/B$.

2.7 The Rate of Change of a Function

Although the idea of the rate of change of a function is best handled within the framework of the calculus and will be discussed from that standpoint in Chapter 5, it will be useful to discuss it briefly from an intuitive point of view at this point.

It seems clear that there is a relationship between the rate at

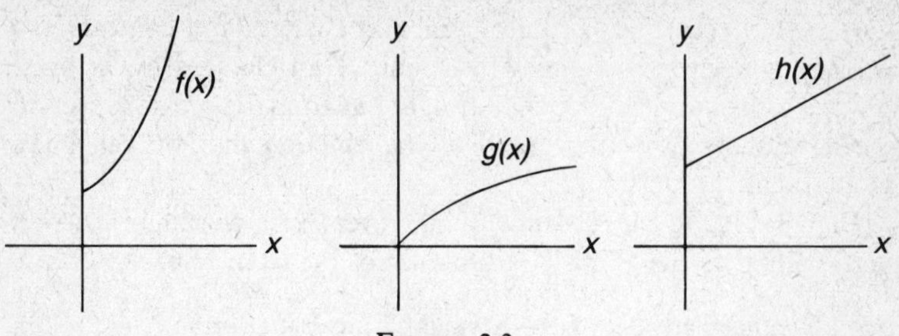

FIGURE 2.3

which a function is increasing (or decreasing) and its graph. Consider the graphs of three functions, $f(x)$, $g(x)$, and $h(x)$, shown in Figure 2.3.

The curve representing $f(x)$ is getting steeper as x increases, or in other words, the function is increasing more and more rapidly. On the other hand, the graph of $g(x)$ appears to be leveling off as x increases, i.e., the rate of increase of the function is getting smaller.

Let us now conceive of a curve representing a function $h(x)$, whose rate of change is constant, i.e., a curve whose *steepness* does not change as we move along the curve. Such a curve corresponds to our idea of a straight line. Accordingly, we shall define a straight line as a curve that is the graph of a function whose rate of change is constant. Only vertical lines are the exception.

2.8 The Rate of Change of a Linear Function

Evidently the values of the function change as x changes. It is a matter of interest and importance to determine the rate at which this change takes place.

Let us consider the change in the function $f(x)$ when x changes by an amount Δx from some specified value x_0, i.e., when x changes from x_0 to $x_0 + \Delta x$. The values of the function for the two specified values of x are clearly $f(x_0) = mx_0 + b$ and $f(x_0 + \Delta x) = m(x_0 + \Delta x) + b$. Hence, the change in the value of the function is given by the difference

$$\begin{aligned} f(x_0 + \Delta x) - f(x_0) &= [m(x_0 + \Delta x) + b] - (mx_0 + b) \\ &= mx_0 + m\Delta x + b - mx_0 - b = m(\Delta x) \end{aligned}$$

However, the rate of change of the function with respect to x is given by the ratio of the change in the function to the change in x. Evidently, this is $m(\Delta x)/\Delta x = m$.

It is thus seen that the linear function changes at a constant rate m, i.e., whenever x changes by 1 unit, $f(x)$ changes by m units. Hence, using our definition of a straight line from the previous section, we have shown that the graph of the function that we call linear is a straight line.

Conversely, it can be shown that every straight line, except a vertical line, is the graph of a function of the form $f(x) = mx + b$.

2.9 Standard Forms of the Straight-Line Equation

We have seen that the graph of the equation $y = mx + b$ is a straight line and that the rate of change of y with respect to x is the constant m. A positive value of m corresponds to a line that rises as x increases, whereas a negative m indicates a line that descends as x increases. A line for which $m = 0$ in horizontal, i.e., parallel to the x axis. Since m is a measure of the steepness of the inclination of a line, it is commonly called the **slope** of the line. Furthermore, since $y = b$ when $x = 0$, the line $y = mx + b$ passes through the point $(0,b)$. Hence, b is commonly called the **y intercept** of the line. In view of these two facts, the equation $y = mx + b$ is usually referred to as the **slope-intercept** form of the straight-line equation. Note also that the slope of a line whose equation is given can be obtained simply by solving the equation for y and observing the coefficient of x. For example, the equation $3x - 5y = 7$ can be written in the form $y = \frac{3}{5}x - \frac{7}{5}$, and the slope of the line is seen to be $\frac{3}{5}$.

Since the slope of a line indicates the amount of change in y per unit change in x, it is clear that if (x_1,y_1) and (x_2,y_2) are any points on the line, the slope can be found by finding the ratio of the change in y between the two points to the change in x between these points. This leads to the formula

$$m = \frac{y_2 - y_1}{x_2 - x_1} \tag{1}$$

where (x_1,y_1) and (x_2,y_2) are any two points on the line and $x_1 \neq x_2$. The exceptional case, i.e., a line for which $y_1 \neq y_2$, but $x_1 = x_2$, is a line parallel to the vertical axis. For such a line the slope is not defined.

If (x,y) is any point on a line and (x_1,y_1) is a specified point, then we have $m = (y - y_1)/(x - x_1)$, which can be written as

$$y - y_1 = m(x - x_1) \tag{2}$$

This is known as the **point-slope** form of the straight-line equation. It can be used to write the equation of a line with a specified slope passing through a specified point.

Example 1 Write in the form $Ax + By + C = 0$ the equation of a line with slope -2 passing through the point $(3,-4)$.
Solution Substituting in the point-slope form of the straight-line equation, we have $y + 4 = -2(x - 3)$, which may be written as $2x + y - 2 = 0$.

Any line that is not parallel to either axis will evidently intersect both the horizontal and vertical axes at exactly one point on each axis. The point of intersection with the y axis has already been designated as the point $(0,b)$. We shall designate the point of intersection of the line with the x axis as the point $(a,0)$. Hence, the slope of the line can be seen to be $m = (b - 0)/(0 - a) = -b/a$. Therefore, employing the point-slope form and using the point $(0,b)$, we have for the equation of the line

$$y - b = \frac{-b}{a}(x - 0)$$

If we divide this equation by b, we have $y/b - 1 = -x/a$. This can be written as

$$\frac{x}{a} + \frac{y}{b} = 1 \tag{3}$$

This is known as the **intercept** form of the straight-line equation. The number a is called the **x intercept** of the line, just as the number b has been designated as the **y intercept.** Thus the intercept form of the straight-line equation enables us to obtain its equation readily if we know its x and y intercepts.

The procedure employed in deriving the intercept form of the straight-line equation can be used to find the equation of a line passing through two specified points.

Example 2 Find the equation of the line passing through the points $(2,-5)$ and $(-4,7)$.
Solution The slope of the required line is clearly $m = (7 + 5)/(-4 - 2) = -2$. Hence, using the point $(2,-5)$ in the point-slope form, we

have $y + 5 = -2(x - 2)$, or $2x + y + 1 = 0$. The student can easily verify the fact that the result would have been the same if the point $(-4,7)$ had been used in the point-slope equation.

None of the forms of the straight-line equation discussed thus far apply to lines parallel to the vertical axis since such lines have no slope and no y intercept. However, this exceptional case is easily taken care of as follows. Consider a line that is parallel to the y axis and that intersects the x axis at the point $(3,0)$. Evidently the x coordinate of every point on this line is 3, regardless of what its y coordinate is. Hence its equation can be written $x = 3$. In the same way the equation of a line parallel to the y axis passing through the point $(-5,0)$ has the equation $x = -5$. Clearly, the equation of any line parallel to the y axis is of the form $x = k$, where k is a real number.

The slopes of two given lines may be used to determine whether they are parallel or perpendicular. It is intuitively evident that two lines with the same slope are parallel provided that the slopes exist. The exceptional case is that of two lines, parallel to the vertical axis, whose slopes are undefined.

It can be shown that two lines are perpendicular to each other if and only if their slopes are negative reciprocals, provided that the slopes of both lines exist. Hence, if the slopes of two lines are m_1 and m_2, respectively, the lines are perpendicular if and only if $m_1 = -1/m_2$. The exceptional case is that in which one line is horizontal and the other vertical. These lines are evidently perpendicular to each other, although the slope of the vertical line does not exist.

In view of the above considerations, it is clear that the lines $3x + 2y - 5 = 0$ and $3x + 2y + 10 = 0$ are parallel, since it is readily seen that the slope of each line is $-\frac{3}{2}$. The lines $3x + 2y - 5 = 0$ and $2x - 3y + 10 = 0$ are perpendicular since the slope of the first line is $-\frac{3}{2}$ while the slope of the second is $\frac{2}{3}$, the slopes being negative reciprocals of each other.

As illustrations of the exceptional cases, the lines $x = 5$ and $x = 10$ are parallel, whereas the lines $x = 5$ and $y = 10$ are perpendicular.

2.10 The Length of a Line Segment

A problem of considerable interest and importance is that of finding the length of a segment of a straight line extending between two

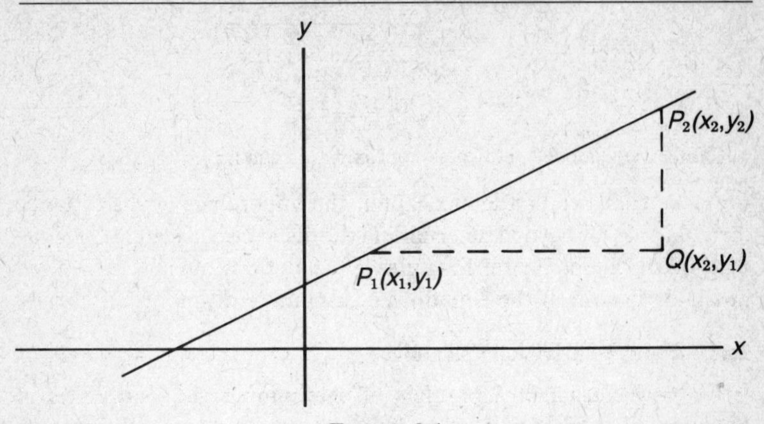

FIGURE 2.4

specified points. This is a special case of the general problem of finding the length of an arc of a curve. The general problem is a rather difficult one, requiring the use of the integral calculus, but in the case of the straight line the problem can be solved by a much more elementary procedure. Figure 2.4 shows a straight line containing two specified points P_1 and P_2 whose coordinates have been designated as (x_1,y_1) and (x_2,y_2), respectively. It is assumed that the same scale of measurement is used on each of the axes.

A line has been drawn through P_1 parallel to the x axis, and a line has been drawn through P_2 parallel to the y axis. The intersection of these two lines has been designated as the point Q. Evidently, its coordinates are (x_2,y_1). Furthermore, the length of the line segment P_1Q is evidently $x_2 - x_1$, and the length of the line segment QP_2 is $y_2 - y_1$. Hence, by the Pythagorean theorem, if the length of P_1P_2 is designated by d, we have

$$d = \sqrt{(x_2 - x_1)^2 + (y_2 - y_1)^2} \tag{1}$$

Formula (1) is usually referred to as the **distance formula** since it enables us to determine the distance between any two points in the plane.

Example 1. Find the distance between the points (2,−4) and (−6,8).
Solution Designating the first point as (x_1,y_1) and the second as (x_2,y_2), we have

$$\begin{aligned} d &= \sqrt{(-6-2)^2 + (8+4)^2} \\ &= \sqrt{64 + 144} = \sqrt{208} \\ &= 4\sqrt{13} \simeq 14.4 \end{aligned}$$

(Note: the symbol $\simeq$ indicates approximate equality.)

As we shall see in a later chapter, the importance of the distance formula goes far beyond determining the distance between two points. One of the fundamental formulas of analytic geometry, it is used in the derivation of the equations of a number of important curves.

2.11 Point of Division Formulas

We now consider the problem of determining the coordinates of the point that divides a given line segment into two subsegments whose lengths have a specified ratio. (See Figure 2.5.)

We wish to determine the point P in such a way that $P_1P/PP_2 = k$, where $0 < k < 1$.

It is readily seen that triangles P_1QP and P_1RP_2 are similar. Therefore,

$$\frac{\overline{P_1P}}{\overline{P_1P_2}} = \frac{\overline{P_1Q}}{\overline{P_1R}} = \frac{x - x_1}{x_2 - x_1} = k$$

Solving for x, we obtain

$$x = x_1 + k(x_2 - x_1)$$

Since triangles PSP_2 and P_1RP_2 are similar,

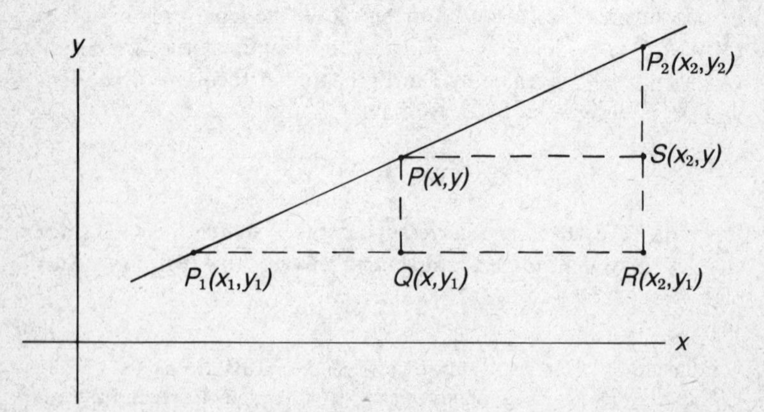

FIGURE 2.5

$$\frac{\overline{P_1P}}{\overline{P_1P_2}} = \frac{\overline{RS}}{\overline{RP_2}} = \frac{y - y_1}{y_2 - y_1} = k$$

Solving for y, we obtain

$$y = y_1 + k(y_2 - y_1) \tag{2}$$

An important special case is the one for which $k = \frac{1}{2}$. In this case the point P bisects the line segment $\overline{P_1P_2}$, and the above formulas are readily seen to reduce to what are called the midpoint formulas:

$$x = \frac{x_1 + x_2}{2} \tag{3}$$

$$y = \frac{y_1 + y_2}{2} \tag{4}$$

Example 1 Find the point which is one-third of the way from the point (2,5) to the point (8,14).

Solution Here $x_1 = 2$, $y_1 = 5$, $x_2 = 8$, $y_2 = 14$, and $k = \frac{1}{3}$. Hence

$$x = 2 + \tfrac{1}{3}(8 - 2) = 4$$

$$y = 5 + \tfrac{1}{3}(14 - 5) = 8$$

Hence the required point is (8,4).

Example 2 Given the points P_1 $(-3,2)$ and P_2 $(7,-3)$, find the point P such that $P_1P/PP_2 = \frac{2}{3}$.

Solution Recalling that k was designated as the ratio P_1P/P_1P_2, we readily see that in this case $k = \frac{2}{5}$. Hence

$$x = -3 + \tfrac{2}{5}(7 + 3) = 1$$
$$y = 2 + \tfrac{2}{5}(-3 - 2) = 0$$

Example 3 Find the midpoint of the line segment connecting the points $(2,-7)$ and $(4,9)$.

Solution Using formulas (3) and (4), we have

$$x = \tfrac{1}{2}(2 + 4) = 3$$
$$y = \tfrac{1}{2}(-7 + 9) = 1$$

2.12 The Quadratic Function and Its Graph

A quadratic function is a function of the form

$$f(x) = ax^2 + bx + c \qquad a \neq 0$$

where $a, b,$ and c are constants and a is not 0. The domain of this function is the set of all real numbers. But what is the range?

The answer to this question is not evident, and we shall supply it below in connection with our discussion of the graph of the quadratic function.

We shall initiate our discussion with a specific example. Consider the function $y = 2x^2 - 8x + 5$. Let us construct a table of corresponding values of x and y, plot the points determined by these ordered pairs, and sketch the curve. The results are shown in Figure 2.6.

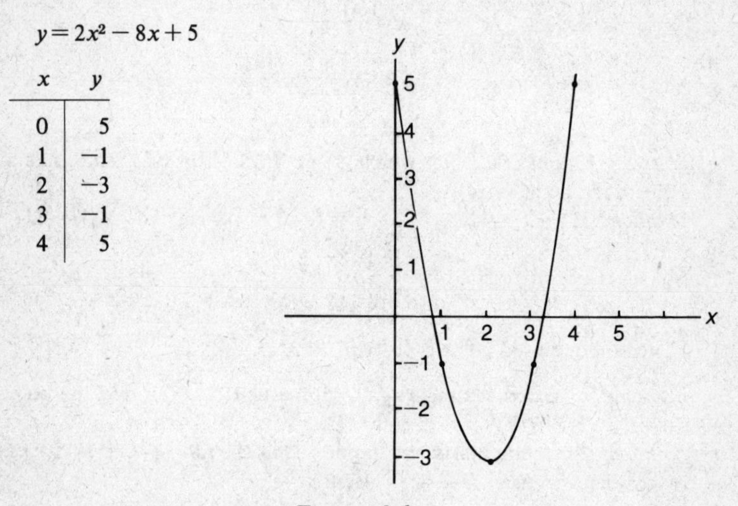

$y = 2x^2 - 8x + 5$

x	y
0	5
1	−1
2	−3
3	−1
4	5

FIGURE 2.6

It appears from a glance at the above graph that the point $(2, -3)$ is the lowest point on the curve. But how can we be sure? The question just posed is equivalent to the one concerning the range of the function that was raised in the preceding paragraph, and we shall now attempt to answer it. In doing so, we shall employ an algebraic technique known as **completing the square.** Since $(x + a)^2 = x^2 + 2ax + a^2$ it is evident that in a perfect square trinomial of this type the last term is the square of half of the coefficient of x. For example, the expression $x^2 + 10x$ can be converted into a perfect square trinomial by the addition of the number 25.

We proceed as follows:

$$\begin{aligned} y &= ax^2 + bx + c \\ &= a\left(x^2 + \frac{b}{a}x + \frac{c}{a}\right) \\ &= a\left(x^2 + \frac{b}{a}x + \frac{b^2}{4a^2} - \frac{b^2}{4a^2} + \frac{c}{a}\right) \end{aligned}$$

where we have added and subtracted $[\frac{1}{2}(b/a)]^2$. Hence

$$\begin{aligned} y &= a\left[\left(x + \frac{b}{2a}\right)^2 - \frac{b^2}{4a^2} + \frac{c}{a}\right] \\ &= a\left[\left(x + \frac{b}{2a}\right)^2 + \frac{4ac - b^2}{4a^2}\right] \end{aligned}$$

The quantity $(x + b/2a)^2$ is non-negative for all x. Its least value occurs when $x = -b/2a$. Hence if $a > 0$, the least value of y occurs when $x = -b/2a$, in which case

$$y = a\left(0 + \frac{4ac - b^2}{4a^2}\right) = \frac{4ac - b^2}{4a}$$

We can therefore state that for $a > 0$ the range of the function $y = ax^2 + bx + c$ is the set

$$\left\{y \mid y \geq \frac{4ac - b^2}{4a}\right\}$$

A similar analysis shows that if $a < 0$, the greatest value of y occurs when $x = -b/2a$ and the range of the function is the set

$$\left\{y \mid y \leq \frac{4ac - b^2}{4a}\right\}$$

It will be recalled that in Figure 2.6 we sketched the graph of the function $y = 2x^2 - 8x + 5$, and the question was raised whether the point $(2,-3)$ was actually the lowest point on the graph. This question can now be answered in the affirmative. In this case $a = 2$, $b = -8$, $c = 5$: from the previous analysis we see that the least value of y, occurring when $x = -b/2a = -8/4 = 2$, is

$$y = \frac{4ac - b^2}{4a} = \frac{(4)(2)(5) - (-8)^2}{(4)(2)} = \frac{-24}{8} = -3$$

Of course, the value $y = -3$ could also be obtained simply by substituting $x = 2$ in the equation $y = 2x^2 - 8x + 5$.

The curve shown in Figure 2.6 is known as a **parabola.** In fact, the graph of any function of the form $y = ax^2 + bx + c$ is a curve of this type. From the previous analysis it should be clear that if $a > 0$, the curve opens upward and has a lowest point; whereas if $a < 0$, the curve opens downward and has a highest point. Using the formulas previously derived, we may easily verify that the highest point on the graph of the function $y = -3x^2 - 18x + 10$ is the point $(-3,37)$. The parabola, a special case of a class of curves known as conics, will be discussed in more detail in a later section of this chapter.

2.13 The Exponential Function

A function of the form $f(x) = a^x$, where a is a positive constant, is called an **exponential function.** The domain of this function is the set of all real numbers. Provided $a \neq 1$, the range is the set of all positive real numbers.

A word is in order, however, with respect to the domain. In our previous discussion of exponents (Section 1.7), the expression a^x was assigned a meaning only for rational values of x. We have, for example, assigned no meaning to the expression $3^{\sqrt{2}}$. However, it is clear that the value of $3^{\sqrt{2}}$ is close to the value of $3^{1.4}$ since $\sqrt{2} \simeq 1.4$. Furthermore, $3^{1.4} = 3^{14/10} = 3^{7/5} = \sqrt[5]{3^7}$. It therefore seems reasonable to make use of the idea of a limit and define $3^{\sqrt{2}}$ as follows:

$$3^{\sqrt{2}} = \lim_{x \to \sqrt{2}} 3^x$$

where it is assumed that x approaches $\sqrt{2}$ through rational values.

The graphs of two exponential functions are shown in Figure 2.7. It is clear from inspection of these graphs that the exponential function is a one-to-one function. It therefore has an inverse function, and this important function will be discussed in the next section.

Exponential functions have important applications in the biological and social sciences. The so-called law of growth, which is exponential in form, is applicable to the growth of bacterial and other populations. A function of exponential type is of great importance in the theory of probability. It will be discussed in Chapter 4.

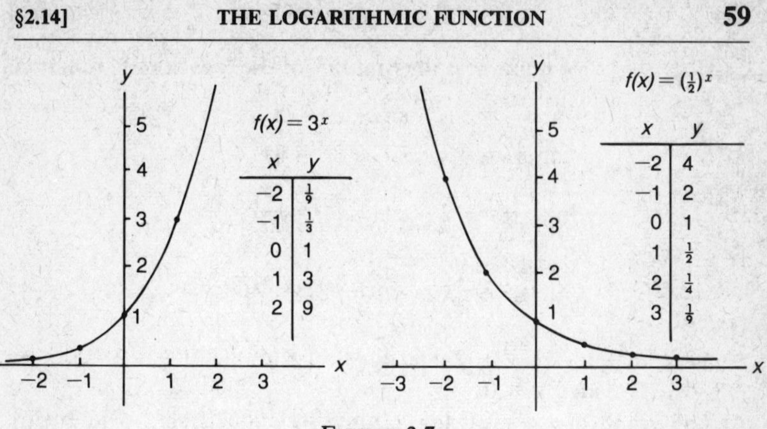

FIGURE 2.7

2.14 The Logarithmic Function

The **logarithmic function** is the inverse of the exponential function. Hence, to obtain this function, we start with the function $f(x) = a^x$ and replace $f(x)$ by y, which gives us the equation $y = a^x$. To obtain the inverse, as indicated in Section 2.5, we interchange x and y to obtain $x = a^y$. Now in order to solve for y in terms of x we must introduce a new notation, which we do by writing

$$y = \log_a x$$

Hence if $f(x) = a^x$, then $f^{-1}(x) = \log_a x$.

From the way in which it has been obtained, it is clear that in connection with the equation $y = \log_a x$ the relationship between x and y is as follows: y is the exponent to which the number a must be raised to equal x. The number a is called the base. It is always a positive number not equal to 1. The equation $y = \log_a x$ is read as "y is the logarithm of x to the base a."

It will be recalled that for the exponential function the domain is the set of all real numbers, and the range is the set of all positive numbers. For the logarithmic function, since it is the inverse of the exponential function, the domain is the set of all positive numbers, and the range is the set of all real numbers. Also since the logarithmic function is one-to-one, no two numbers have the same logarithm.

Here are some numerical illustrations of the logarithmic function.

$$\log_2 8 = 3 \qquad \text{since } 2^3 = 8$$

$$\log_3 81 = 4 \qquad \text{since } 3^4 = 81$$

$$\log_4 \frac{1}{16} = -2 \qquad \text{since } 4^{-2} = \frac{1}{4^2} = \frac{1}{16}$$

$$\log_8 4 = \frac{2}{3} \qquad \text{since } 8^{2/3} = \sqrt[3]{64} = 4$$

Note that $\log_a a = 1$, since $a^1 = a$, and that $\log_a 1 = 0$, since $a^0 = 1$, provided $a \neq 0$.

The graph of a typical logarithmic function is shown in Figure 2.8.

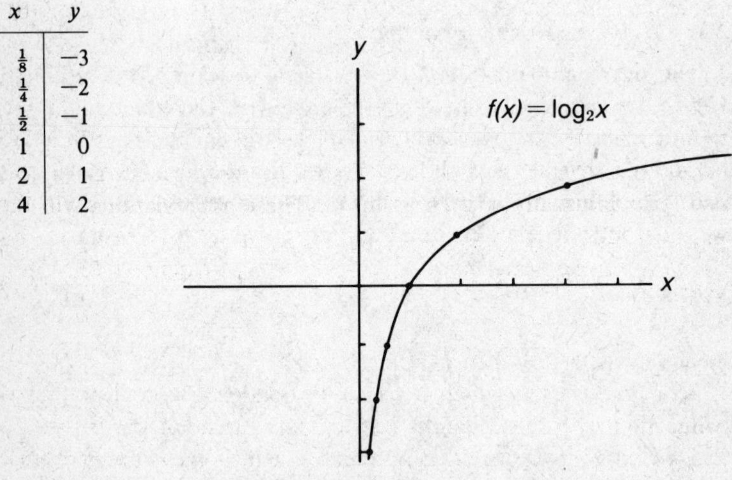

x	y
$\frac{1}{8}$	-3
$\frac{1}{4}$	-2
$\frac{1}{2}$	-1
1	0
2	1
4	2

FIGURE 2.8

The logarithmic function may be shown to have the following important properties. Here x and y are assumed to be positive real numbers and p is any real number.

(1) $\log_a(xy) = \log_a x + \log_a y$

(2) $\log_a \left(\frac{x}{y}\right) = \log_a x - \log_a y$

(3) $\log_a x^p = p \log_a x$

Although a detailed proof of these laws will not be given, it may be noted that a logarithm is essentially an exponent and that these properties follow from the laws of exponents. For example, in (1) multiplication is replaced by addition just as exponents are added when exponential expressions are multiplied.

Although any positive real number except the number 1 may be used as a base for a system of logarithms, there are only two bases in common use. These are the number 10 and a number designated by the letter e, which is an irrational number and for which an exact value cannot be given. An approximation to e, correct to four decimal places, is 2.7183.

Base 10 logarithms are called **common logarithms.** The number 10 is a very convenient base since it is also the base of our counting system. Common logarithms are used extensively for computational purposes. These procedures are discussed in many elementary algebra and trigonmetry books.

Base e logarithms are called **natural logarithms.** They are of great importance in the calculus and will be discussed in Chapter 5.

It is customary to abbreviate $\log_{10} x$ by writing simply $\log x$. Also $\log_e x$ is usually written as $\ln x$. These abbreviations will be used in this book.

Here are some examples that illustrate the properties of the logarithmic function.

Example 1 Solve the equation

$$\log_3 x = 4$$

Solution From the definition of a logarithm it is clear that the given equation is equivalent to

$$x = 3^4 = 81$$

Example 2 Solve the equation

$$\log_x \left(\frac{1}{125}\right) = -3$$

Solution This equation is equivalent to

$$x^{-3} = \frac{1}{125}$$

Hence
$$\frac{1}{x^3}=\frac{1}{125}$$
$$x^3=125$$
$$x=\sqrt[3]{125}=5$$

Example 3 Solve the equation

$$\log x+\log(x-4)=\log 12$$

Solution Using property (1), we obtain

$$\log[x(x-4)]=\log 12$$

Hence
$$x(x-4)=12$$
$$x^2-4x=12$$
$$x^2-4x-12=0$$
$$(x+2)(x-6)=0$$
$$x=-2,\ x=6$$

However, -2 is not an element of the solution set since neither term on the left-hand side of the original equation is defined for that value of x. The number 6 is readily seen to be the only element of the solution set.

Example 4 Solve the equation

$$\log_3 x-\log_3(x-1)=2$$

Solution Using property (2), we obtain

$$\log_3\left(\frac{x}{x-1}\right)=2$$

Hence
$$\frac{x}{x-1}=3^2=9$$
$$x=9x-9$$
$$-8x=-9$$
$$x=\tfrac{9}{8}$$

The number $\frac{9}{8}$ is easily seen to be an element of the solution set. Hence it is the only element.

Example 5 Solve the equation

$$\log\sqrt{x+4}=1.5$$

Solution Since $\sqrt{x+4}=(x+4)^{1/2}$, we obtain, using property (3),

$$\tfrac{1}{2}\log(x+4) = 1.5$$
$$\log(x+4) = 3$$
$$x+4 = 10^3 = 1000$$
$$x = 996$$

This may be verified to be the only element of the solution set.

Logarithms are very useful in solving equations containing exponential expressions, as indicated in the following example.

Example 6 Solve the equation

$$3^x = 15$$

Solution Taking the common logarithm of both sides of the equation, we have

$$x \log 3 = \log 15$$

or

$$x = \frac{\log 15}{\log 3} \simeq \frac{1.1761}{0.4771} \simeq 2.47$$

The logarithms above have been obtained from a table of common logarithms.

2.15 Analytic Geometry

In Section 2.2 we explored the relationship between functions and plane curves that were referred to as the graphs of functions. More generally, we may speak of the graph of an equation even though the equation may not define a function. Consider, for example, the equation

$$x^2 - y^2 = 16$$

If we solve this equation for y, we obtain

$$y = \pm\sqrt{x^2 - 16}$$

If we solve for x, we obtain

$$x = \pm\sqrt{y^2 + 16}$$

Clearly, neither of these last two equations defines a function since in a function, to each element of the domain there corresponds one and only one element of the range. Nevertheless, it is also clear that, using either of these equations, we could obtain a table of corresponding values of x and y and use these ordered pairs to construct a graph, which would be referred to as the graph of the equation $x^2 - y^2 = 16$. Note that since $x^2 - y^2 = 16$ defines a relation, we could also refer to it as the graph of a relation.

This connection between equations and graphs forms the basis for the study of **analytic geometry,** i.e., the study of geometry by algebraic methods.

The graphs of linear functions and therefore of linear equations in two variables have already been discussed (Section 2.2), since every such linear equation defines a linear function. We now proceed to discuss the graphs of quadratic, or second-degree equations, in two variables. The most general quadratic equation in two variables is one of the form

$$Ax^2 + By^2 + Cxy + Dx + Ey + F = 0 \tag{1}$$

where A, B, C, D, E, and F are constants and A, B, and C are not all equal to 0. The graphs of all equations of this type are, except for certain exceptional cases, members of a family of curves called **conics.** There are four basic types of conics: circles, parabolas, ellipses, and hyperbolas. These curves are called conics because each of them is the intersection of a right circular cone and a plane, the particular type of curve depending on the direction in which the section, or slice of the cone, is made.

Our discussion of these curves can be simplified somewhat by using the special case of Equation (1), for which $C = 0$. We shall therefore use as our general equation

$$Ax^2 + By^2 + Dx + Ey + F = 0 \tag{2}$$

where A and B are not both equal to 0.

Each of these curves may be defined in terms of distances. In deriving their standard equations we shall make use of the distance formula obtained in Section 2.10. It is the most important formula in analytic geometry, and we repeat it here. The distance between two points (x_1,y_1) and (x_2,y_2) is given by

$$d = \sqrt{(x_2 - x_1)^2 + (y_2 - y_1)^2} \tag{3}$$

2.16 The Circle

A **circle** may be defined as the set of all points equidistant from a fixed point. If the fixed point is designated as (h,k), the common distance as r, and any point of the circle as (x,y), we have, using formula (3),

$$r = \sqrt{(x-h)^2 + (y-k)^2}$$

or
$$(x-h)^2 + (y+k)^2 = r^2 \tag{1}$$

Equation (1) is usually called the standard equation of a circle with center at (h, k) and radius r. For example, the equation of the circle with center at $(2, -5)$ and radius 6 may be written as

$$(x-2)^2 + (y+5)^2 = 36$$

or
$$x^2 + y^2 - 4x + 10y - 7 = 0$$

Note that this last equation is of the form of Equation (2), Section 2.15, with $A = B = 1$, $D = -4$, $E = 10$, and $F = -7$. The graph of this equation is shown in Figure 2.9.

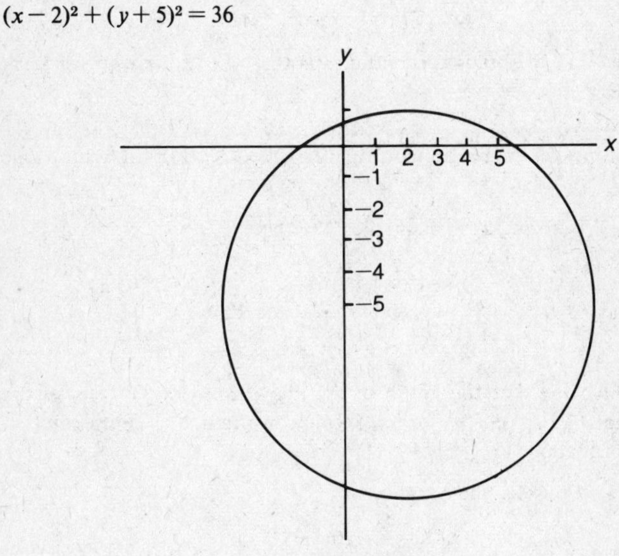

FIGURE 2.9

It may easily be shown that the graph of every equation of the form of Equation (2) in which $A = B$ is either a circle, a single point, or consists of no points whatsoever. Examples of these three cases follow.

Example 1 Consider the equation

$$x^2 + y^2 + 4x - 6y - 3 = 0$$

Solution We shall now transform this equation into the standard form using the previously discussed technique of completing the square. First, rearranging the terms of the equation, we have

$$x^2 + 4x + y^2 - 6y = 3$$

We now use the technique of completing the square described in Section 2.12 to convert both the terms in x and the terms in y into perfect square trinomials. We then have

$$x^2 + 4x + 4 + y^2 - 6y + 9 = 3 + 13$$

where we have added 13 to the right-hand side of the equation to balance the terms we have added on the left. This last equation can now be written as

$$(x + 2)^2 + (y - 3)^2 = 16$$

and we see that the original equation was that of a circle with center at $(-2,3)$ and radius 4.

Example 2 Consider the equation

$$x^2 + y^2 - 2x - 6y + 10 = 0$$

Solution Proceeding as in Example 1, we have

$$x^2 - 2x + y^2 - 6y = -10$$
$$x^2 - 2x + 1 + y^2 - 6y + 9 = -10 + 10 = 0$$
$$(x - 1)^2 + (y - 3)^2 = 0$$

It is readily seen that the graph of the above equation consists of the single point $(1,3)$ since this ordered pair constitutes the entire solution set of the equation.

Example 3 Consider the equation

$$x^2 + y^2 + 8x + 10y + 51 = 0$$

Solution Proceeding as before, we have

$$x^2 + 8x + y^2 + 10y = -51$$
$$x^2 + 8x + 16 + y^2 + 10y + 25 = -51 + 41 = -10$$
$$(x + 4)^2 + (y + 5)^2 = -10.$$

Since the square of any real number is non-negative, it is evident that the solution set of the above equation is empty, and hence its graph does not exist.

Example 4 Consider the equation

$$4x^2 + 4y^2 - 4x + 16y - 19 = 0$$

Solution Note that in this equation $A = B = 4$. Before employing the technique of completing the square, it is necessary to divide each term of the equation by 4 so that the coefficients of x^2 and y^2 will each be 1. If we do this, and at the same time rearrange the terms, we have

$$x^2 - x + y^2 + 4y = \frac{19}{4}$$

$$x^2 - x + \frac{1}{4} + y^2 + 4y + 4 = \frac{19}{4} + \frac{17}{4} = \frac{36}{4} = 9$$

$$\left(x - \frac{1}{2}\right)^2 + (y + 2)^2 = 9$$

This is evidently the equation of a circle with center at $(\frac{1}{2}, -2)$ and radius 3.

The circle is one of the most widely used of all curves because of its simplicity and symmetry. It is frequently employed in the design of buildings, streets, parks, and in many other areas of artistic endeavor. One of its important properties is that it has constant curvature; i.e., a tangent to the curve at a point rotates at a constant rate as the point moves at a constant rate around the curve. It is this fact that makes the circular shape suitable for the wheel, without which our present state of civilization could not have been attained.

2.17 The Ellipse

An **ellipse** may be defined as the set of all points, the sum of whose distances from two fixed points is a constant. The two fixed points are called the **foci** of the ellipse. If we choose as the foci the points $(c,0)$ and $(-c,0)$ and designate a point on the ellipse as (x,y) and the constant sum of the distances as $2a$ (a and c are assumed to be positive), we have

$$\sqrt{(x-c)^2 + y^2} + \sqrt{(x+c)^2 + y^2} = 2a \qquad (1)$$

Since the algebraic process of simplifying this equation is quite lengthy, it will be summarized here, with the details left as an exercise for the student. Briefly, we transpose one of the radical terms, square both sides of the equation, do some more transposing to isolate the radical term that remains, and then square again. The resulting equation will then have the form

$$(a^2 - c^2)x^2 + a^2y^2 = a^2(a^2 - c^2) \qquad (2)$$

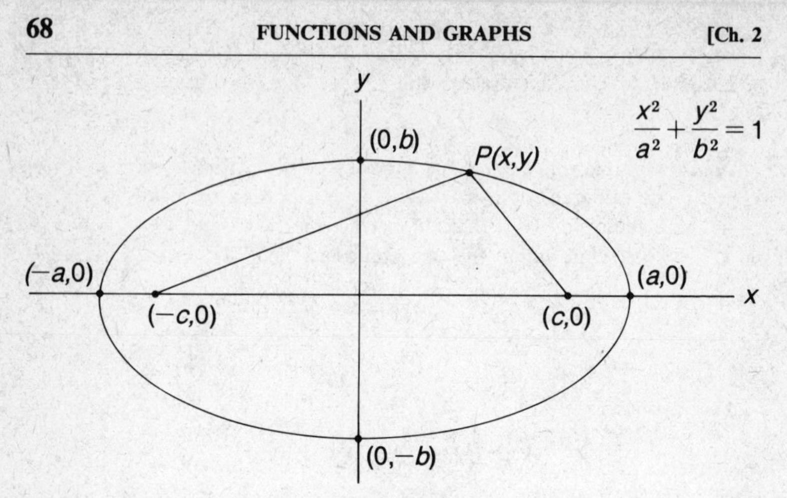

FIGURE 2.10

The distance between the two fixed points is $2c$, and it is clear from the definition of the ellipse that $2a > 2c$, or $a > c$ (see Figure 2.10).* Hence $a^2 - c^2$ is a positive number, and it is therefore appropriate to replace it by the symbol b^2 ($b > 0$). If we do this, the equation of the ellipse may be reduced to the standard form:

$$\frac{x^2}{a^2} + \frac{y^2}{b^2} = 1 \tag{3}$$

We note that if $y = 0$, Equation (3) reduces to $x^2 = a^2$, or $x = \pm a$. Hence the x intercepts are at $(a,0)$ and $(-a,0)$. Similarly, the y intercepts are seen to be $(0,b)$ and $(0,-b)$. Clearly, in view of the relationship $a^2 - c^2 = b^2$, it follows that $a > b$.

The segment of the x axis between the x intercepts, evidently of length $2a$, is called the **major axis** of the ellipse. Similarly, the segment of the y axis between the y intercepts, of length $2b$, is called the **minor axis.** The intersection of the major and minor axes, which is, of course, the origin in this case, is called the **center** of the ellipse. Clearly if $a = b$, Equation (3) can be written in the form $x^2 + y^2 = a^2$, which is evidently the equation of a circle, with center at the origin and radius a. Hence a circle may be regarded as a limiting case of an ellipse in the sense that it is the curve approached by

* This follows from the fact that the sum of the lengths of two sides of a triangle is greater than the length of the third side.

the ellipse as the lengths of the major and minor axes approach equality.

It will be recalled that a and b are related by the equation $b^2 = a^2 - c^2$. Clearly, a value of c close to 0 implies that a and b are nearly equal, i.e., the ellipse is almost circular. On the other hand if c is close to a in value, then a is considerably larger than b, and the ellipse is more or less cigar-shaped. For this reason the quantity c/a, called the **eccentricity,** is a measure of the shape of the ellipse. An ellipse with eccentricity close to 0 is almost circular, whereas an ellipse with eccentricity close to 1 is long and thin.

Returning to Equation (2), Section 2.15, it can be shown that if $A \neq B$ and if A and B have the same sign, then the equation will be that of an ellipse, or in exceptional cases (as was the case with the circle), the graph will consist of a single point or there will be no graph whatsoever.

Example 1 Consider the equation

$$9x^2 + 16y^2 - 144 = 0$$

Solution Dividing each term of the equation by 144 and transposing the constant term, we obtain

$$\frac{x^2}{16} + \frac{y^2}{9} = 1$$

which is readily seen to be in the standard form for the equation of an ellipse with center at the origin. Furthermore, we observe that $a = 4$ and $b = 3$, so that the major axis is of length 8 and the minor axis is of length 6. Also, since $b^2 = a^2 - c^2$ implies $c^2 = a^2 - b^2$, we obtain $c^2 = 16 - 9$, or $c = \sqrt{7} \simeq 2.65$. Hence the eccentricity $= c/a \simeq 2.65/4 \simeq 0.66$.

Example 2 Consider the equation

$$16x^2 + 25y^2 - 64x + 150y - 111 = 0$$

Solution This equation can be written in the form

$$16(x^2 - 4x) + 25(y^2 + 6y) = 111$$

Completing the square yields

$$16(x^2 - 4x + 4) + 25(y^2 + 6y + 9) = 111 + 289 = 400$$

or

$$\frac{(x-2)^2}{25} + \frac{(y+3)^2}{16} = 1 \tag{4}$$

It will be noted that this equation is similar in form to the standard equation of the ellipse. In fact, if we were to let $x - 2 = X$ and $y + 3 = Y$, we would have the standard form $X^2/25 + Y^2/16 = 1$ of an ellipse with center at the origin. Now $X = 0$ implies $x = 2$ and $Y = 0$ implies $y = -3$. Hence, it might be surmised that Equation (4) above is the equation of an ellipse with center at (2,−3). A more detailed analysis proves this surmise to be correct. Furthermore, it may be shown that the length of the axes and the eccentricity may be calculated in the same way as would be done if the center were at the origin, i.e., if the equation were $x^2/25 + y^2/16 = 1$. In this case, the major axis is of length 10, the minor axis is of length 8, and since $c^2 = 25 - 16 = 9$ or $c = 3$, the eccentricity is $c/a = 3/5 = 0.6$.

The ellipse is a curve of considerable practical interest since the orbits of all of the planets around the sun are ellipses as are the orbits around the earth and the moon of the space vehicles carrying astronauts. The orbit of the earth around the sun is an ellipse with eccentricity approximately equal to 0.17, a relatively small value, which indicates that the orbit is nearly circular.

An interesting phenomenon associated with ellipses is the existence of buildings that have the **whispering gallery** property. Such buildings have ellipsoidal ceilings, an ellipsoid being a surface generated by rotating an ellipse about one of its axes. The points corresponding to the foci of the generating ellipse are usually marked in the floor of the building. A low whisper emitted by a person standing at one of the foci can be readily heard by a person standing at the other even though it may be completely inaudible to other persons standing nearby. This is due to the fact that all sound waves projected from one of the foci are reflected by the ceiling to the other. See Figure 2.11, in which F_1 and F_2 are the foci.

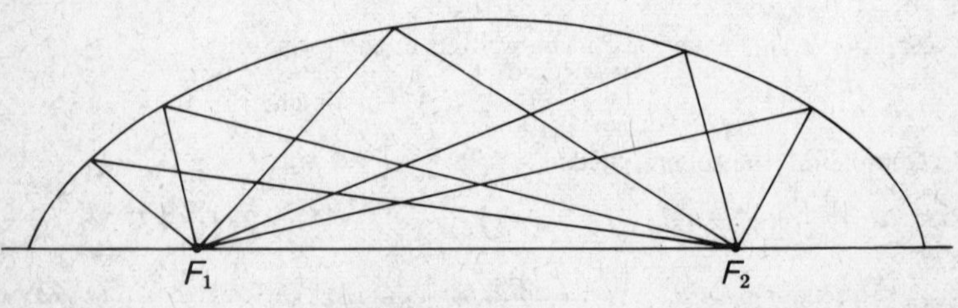

FIGURE 2.11

2.18 The Hyperbola

The definition of the curve known as the **hyperbola** may be obtained by substituting "absolute value of the difference" for "sum" in the definition of the ellipse. In other words, a hyperbola is the set of all points, the absolute value of the difference of whose distances from two fixed points is a constant. As in the case of the ellipse, the two fixed points are called the **foci.** In deriving the equation of the curve we shall choose them at $(c,0)$ and $(-c,0)$ and take the absolute value of the difference as $2a$, where a and c are positive:

$$\sqrt{(x-c)^2+y^2}-\sqrt{(x+c)^2+y^2}=\pm 2a$$

We now proceed to simplify this equation in exactly the same way as we did in the case of the ellipse, and we obtain

$$(c^2-a^2)x^2-a^2y^2=a^2(c^2-a^2)$$

In this case it is clear that $2c > 2a$, or $c > a$. (See Figure 2.12.*) Hence $c^2 - a^2$ is positive and it is appropriate to replace it by b^2 $(b > 0)$. The above equation may then be written in the form

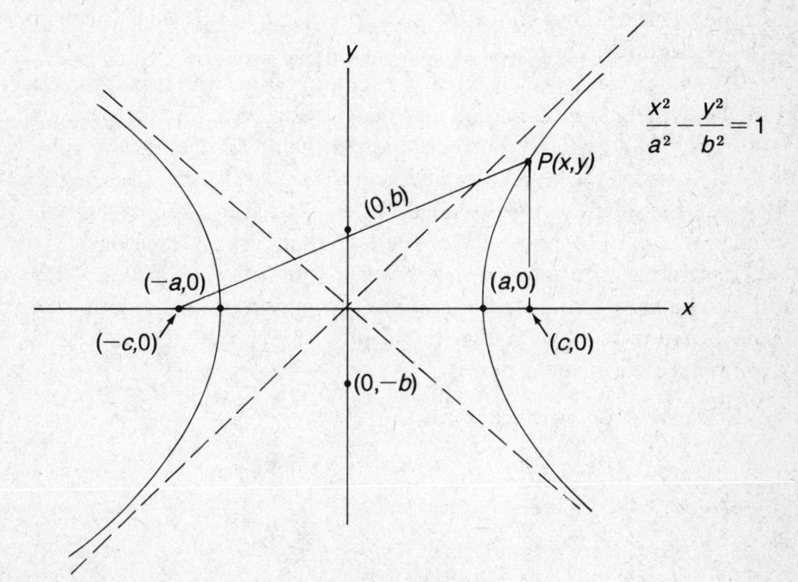

FIGURE 2.12

* This follows from the fact that the absolute value of the difference of two sides of a triangle is less than the length of the third side.

$$\frac{x^2}{a^2} - \frac{y^2}{b^2} = 1 \tag{1}$$

Setting $y = 0$, we note that the x intercepts of the curve are at $(a,0)$ and $(-a,0)$. If $x = 0$, we obtain $y^2 = -b^2$, which is satisfied only by imaginary values of y. Hence there are no y intercepts, as is indicated by the graph in Figure 2.12. In fact, if we solve Equation (1) for y, we obtain

$$y = \pm\frac{b}{a}\sqrt{x^2 - a^2} \tag{2}$$

From Equation (2), it is clear that y is real only if $x \geq a$ or $x \leq -a$. Hence, there are no points of the curve in the interval $-a < x < a$, and the hyperbola therefore consists of two completely unconnected branches.

The segment of the x axis between $(-a,0)$ and $(a,0)$, called the **transverse axis** of the hyperbola, is evidently of length $2a$. The segment of the y axis between the points $(0,-b)$ and $(0,b)$ is called the **conjugate axis,** and its length is of course $2b$. The intersection of the transverse and conjugate axes (the origin in this case) is called the **center** of the hyperbola.

It can be shown that the lines $y = \pm(b/a)x$ frame the two branches of the hyperbola in the sense that these branches approach the lines more and more closely but never touch them. These lines are the asymptotes of the hyperbola (Section 2.2). As in the case of the ellipse, the quantity c/a is called the eccentricity of the hyperbola. Since $c > a$, the eccentricity of the hyperbola always exceeds 1.

Returning once more to the general equation (2), Section 2.15, it can be shown that if A and B are of opposite sign, the graph of the equation is either a hyperbola, or in exceptional cases, a pair of intersecting straight lines.

Example 1 Consider the equation

$$4x^2 - 9y^2 - 32x - 36y - 8 = 0$$

Solution Proceeding as before, we have

$$4(x^2 - 8x) - 9(y^2 + 4y) = 8$$

$$4(x^2 - 8x + 16) - 9(y^2 + 4y + 4) = 8 + 28 = 36$$

$$\frac{(x-4)^2}{9} - \frac{(y+2)^2}{4} = 1$$

Using the same reasoning as was employed in the case of the ellipse, we can see that this is the equation of a hyperbola with center at $(4,-2)$.

A hyperbola for which $a = b$ is called an **equilateral hyperbola.** It may be shown that the graph of an equation of the form $xy = k$, where k is a constant, is an equilateral hyperbola in which the coordinate axes are asymptotes. It will be recalled from Chapter 1 that y is said to vary inversely with x if the relationship $y = k/x$ holds for all values of x and y. Hence the graph of the function defining inverse variation is a hyperbola.

2.19 The Parabola

A **parabola** may be defined as the set of all points equidistant from a fixed point and a fixed straight line. The fixed point is called the **focus,** and the fixed straight line is called the **directrix.**

We choose the point $(0,c)$ as the focus and the line $y = -c$ as the directrix. (See Figure 2.13.) If $c > 0$, the focus will be above the origin, as shown in Figure 2.13. If (x,y) is a point of the curve, we have, by the distance formula,

$$\sqrt{x^2 + (y - c)^2} = y + c$$

Squaring both sides yields

$$x^2 + y^2 - 2cy + c^2 = y^2 + 2cy + c^2$$

Hence

$$x^2 = 4cy \tag{1}$$

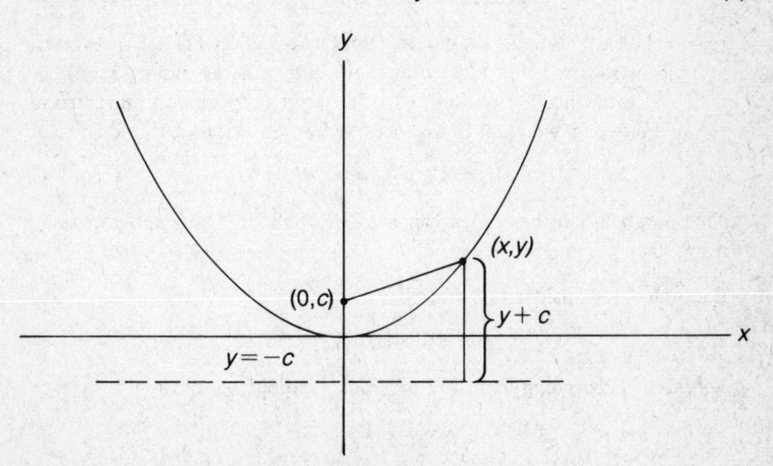

FIGURE 2.13

If this equation is written in the form $y = [1/4c]x^2$, it is readily seen that $y \geq 0$ and that the lowest point on the curve is the origin. The point midway between the focus and the directrix, the origin in this case, is called the **vertex** of the parabola. The graph is shown in Figure 2.13.

If $c < 0$, the focus will be below the origin. However, the same equation will be obtained with $y \leq 0$, the origin the highest point of the curve, and the parabola opening downward.

It will be recalled that in Section 2.12, it was pointed out that the graph of an equation of the form $y = ax^2 + bx + c$ was called a parabola and that the parabola opens upward if $a > 0$ and downward if $a < 0$. It is readily seen that Equation (1) is a special case of an equation of this type, in which a has the value $1/4c$ and in which b and c are equal to 0.

Returning to Equation (2), Section 2.15, we can show that the graph of this equation will be a parabola if either A or B is 0, but not if both are.

Example 1 Consider the equation

$$x^2 - 4x - 6y - 32 = 0$$

Solution This equation may be transformed as follows:

$$x^2 - 4x = 6y + 32$$
$$x^2 - 4x + 4 = 6y + 32 + 4 = 6y + 36$$
$$(x - 2)^2 = 6(y + 6)$$

The last equation is similar in form to equation (1). However the vertex, instead by being located at the origin, is at the point $(2,-6)$.

If the focus had been chosen as the point $(c,0)$ and the directrix as the line $x = -c$, we would have obtained an equation of the form

$$y^2 = 4cx \qquad (2)$$

This parabola can be seen to open to the right if $c > 0$ and to the left if $c < 0$.

Example 2 Consider the equation

$$y^2 + 2y + 4x - 11 = 0$$

Solution Proceeding as in the previous example, we have

$$y^2 + 2y = -4x + 11$$
$$y^2 + 2y + 1 = -4x + 11 + 1 = -4x + 12$$
$$(y + 1)^2 = -4(x - 3)$$

This equation is similar in form to Equation (2). The vertex is at the point $(3,-1)$, and the parabola opens to the left.

2.20 Systems of Quadratic Equations

In this section we shall discuss, by means of examples, systems of equations consisting of either one linear and one quadratic equation or of two quadratic equations. As will be seen, the procedures employed will involve elimination of one variable by substitution.

Example 1 Consider the system

$$x^2 + 4y^2 = 25 \tag{1}$$

$$3x - 2y = 5 \tag{2}$$

Solution Solving for x in Equation (2), we have $x = (2y + 5)/3$. Substituting in Equation (1), we obtain

$$[(2y+5)/3]^2 + 4y^2 = 25$$
$$(4y^2 + 20y + 25)/9 + 4y^2 = 25$$
$$4y^2 + 20y + 25 + 36y^2 = 225$$
$$40y^2 + 20y - 200 = 0$$
$$2y^2 + y - 10 = 0$$
$$(y-2)(2y+5) = 0$$
$$y = 2,\ y = -5/2$$

Substituting $y = 2$ in the equation $x = (2y + 5)/3$, we find $x = 3$. In the same way we see that if $y = -\frac{5}{2}$, then $x = 0$. Hence the solution set consists of the two ordered pairs $(3,2)$ and $(0,-\frac{5}{2})$. From the graphical standpoint, we have found the points of intersection of an ellipse and a straight line. See Figure 2.14.

Example 2 Consider the system

$$x^2 + y^2 = 17 \tag{3}$$

$$y = 4x^2 \tag{4}$$

Solution From (4) we have $x^2 = y/4$. Substituting in (3), we obtain

$$y/4 + y^2 = 17$$

or

$$y + 4y^2 = 68$$
$$4y^2 + y - 68 = 0$$
$$(y-4)(4y+17) = 0$$
$$y = 4,\ y = -\tfrac{17}{4}$$

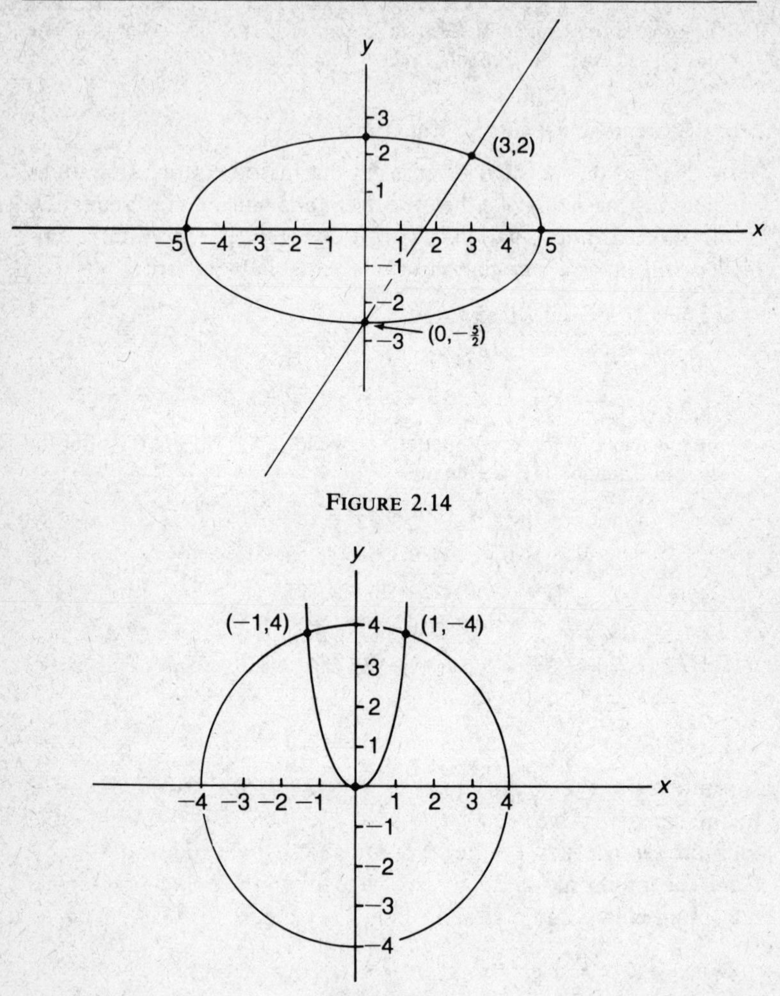

FIGURE 2.14

FIGURE 2.15

Since $x^2 = y/4$, when we substitute $y = 4$ we obtain $x^2 = 1$, or $x = \pm 1$. Clearly the substitution of $y = -\frac{17}{4}$ results in imaginary values of x, so that if we restrict our solution set to ordered pairs of real numbers, it consists of

$$(1,4) \text{ and } (-1,4)$$

As indicated in Figure 2.15, we have obtained the points of intersection of a circle and a parabola.

Example 3 Consider the system

$$4x^2 + y^2 = 32 \tag{5}$$

$$x^2 - y^2 = -12 \tag{6}$$

Solution If Equations (5) and (6) are added, we obtain

$$5x^2 = 20$$
$$x^2 = 4$$
$$x = \pm 2$$

From Equation (5) we have

$$y^2 = x^2 + 12$$

Substituting $x = \pm 2$, we obtain

$$y^2 = 16$$
$$y = \pm 4$$

The solution set therefore consists of the four ordered pairs

$$(2,4),\ (2,-4),\ (-2,4),\ (-2,-4)$$

As indicated in Figure 2.16, we have obtained the four points of intersection of an ellipse and a hyperbola. In this case the foci of both curves are on the y axis.

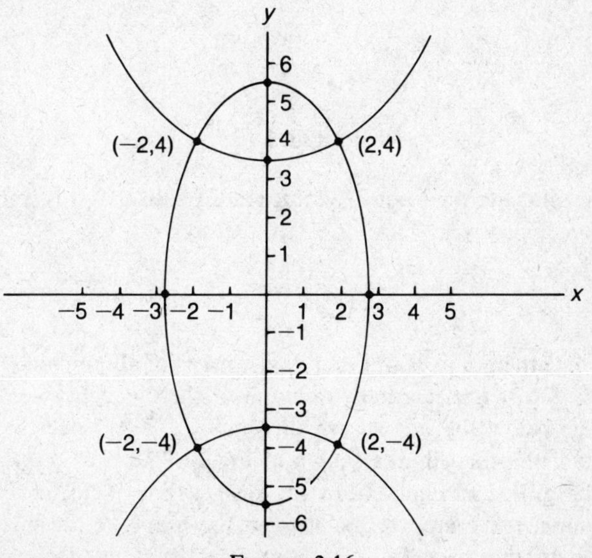

FIGURE 2.16

2.21 Graphs of Linear Inequalities and Linear Programming

Linear inequalities in two variables may be readily graphed on a Cartesian coordinate system. Consider, for example, the inequality

$$2x + 3y < 12$$

In graphing this inequality, it will be helpful to start with the graph of the linear equation

$$2x + 3y = 12$$

The graph of this equation consists of all points on the straight line shown in Figure 2.17.

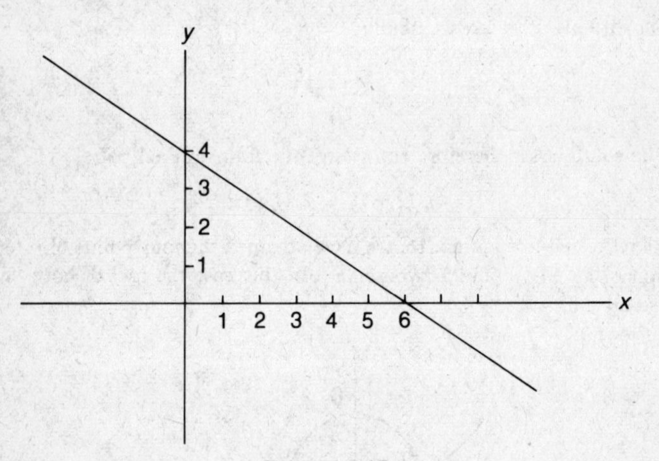

FIGURE 2.17

Now, to return to the inequality, it is readily seen that it is equivalent to the inequality

$$y < 4 - \frac{2}{3}x$$

Now, since the graph of the equation consists of all points for which $y = 4 - \frac{2}{3}x$, it is clear that the graph of the inequality consists of all points below the line shown in Figure 2.17. Hence the graph consists of the shaded area shown in Figure 2.18.

In the graph in Figure 2.18 the dashed line indicates that the points on the line are not included in the graph of the inequality. However, if the inequality had been $2x + 3y \leq 12$, the points on

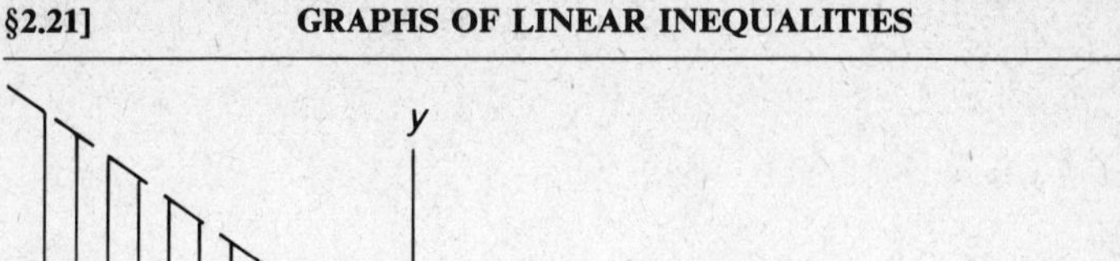

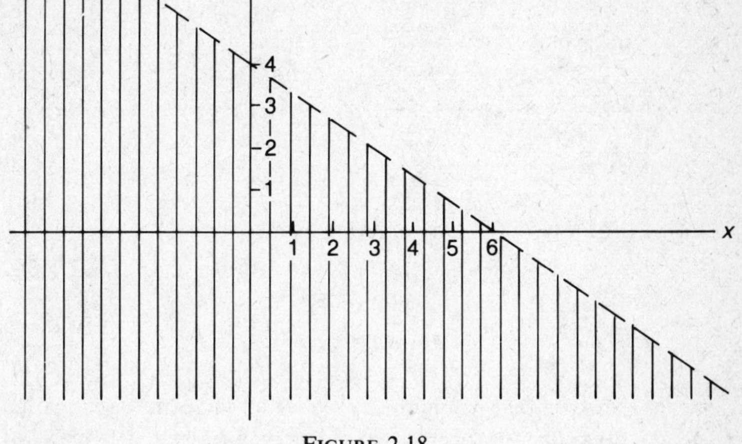

FIGURE 2.18

the line would have been included in the graph and a solid line would have been drawn.

We proceed now from the graph of a single inequality to the graph of a system of several inequalities. Consider the following system:

$$3x + 4y \leq 10 \qquad (1)$$

$$x - 5y \geq 16 \qquad (2)$$

$$x \geq -2 \qquad (3)$$

It is readily seen that the graph of (1) consists of all points below or on the line $3x + 4y = 10$. The graph of (2) consists of all of the points above or on the line $x - 5y = 16$. Finally, the graph of (3) consists of all the points on or to the right of the line $x = -2$. If the three lines referred to above are graphed on the same coordinate system, they form a triangle. It follows from the previous discussion that the graph of the system of inequalities consists of all points inside or on the boundaries of the triangle. (See Figure 2.19.) In other words, the coordinates of each of the points in this region satisfy each of the three inequalities above.

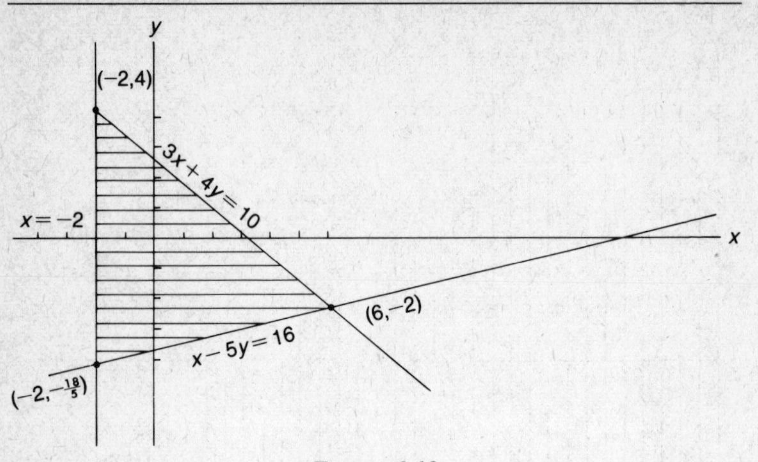

FIGURE 2.19

Graphs such as that in Figure 2.19 have an important application to **linear programming** problems, which may be stated as follows. Given a linear function of n variables* $x_1, x_2, \ldots, x_n$ subject to a system of linear inequalities, find a set of values for the variables that maximizes or minimizes the function. We shall illustrate this type of problem with the following example, involving only two variables.

An automobile manufacturer makes two kinds of cars, a compact called a *Scrambler* and a luxury car called a *Roadroyal.* Because of transportation limitations, the total number of cars manufactured each week cannot exceed 100. It takes 100 person-hours to manufacture a *Scrambler* and 300 person-hours to manufacture a *Roadroyal.* A union contract requires that the total number of person-hours must be at least 12,000. If the profit on a *Scrambler* is \$400 and the profit on a *Roadroyal* is \$300, how many of each kind of car should be manufactured weekly in order to maximize the total profit?

If the number of *Scramblers* is denoted by x_1 and the number of *Roadroyals* by x_2, then it is readily seen that the limitations stated above may be expressed in the form of the following inequalities.

(1) $x_1 \geq 0$

(2) $x_2 \geq 0$

* A function such as $f(x,y) = 2x + 3y + 5$. Functions of several variables will be formally defined in Section 5.23.

(3) $x_1 + x_2 \leq 100$

(4) $100x_1 + 300x_2 \geq 12{,}000$ or $x_1 + 3x_2 \geq 120$

Furthermore, the function to be maximized is

$$f(x_1,x_2) = 400x_1 + 300x_2$$

The graph of inequalities (1) through (4) is shown in Figure 2.20. The set of points in the shaded area is called a **polyhedral convex set,** and the points at the intersections of the bounding lines are called **corner points.**

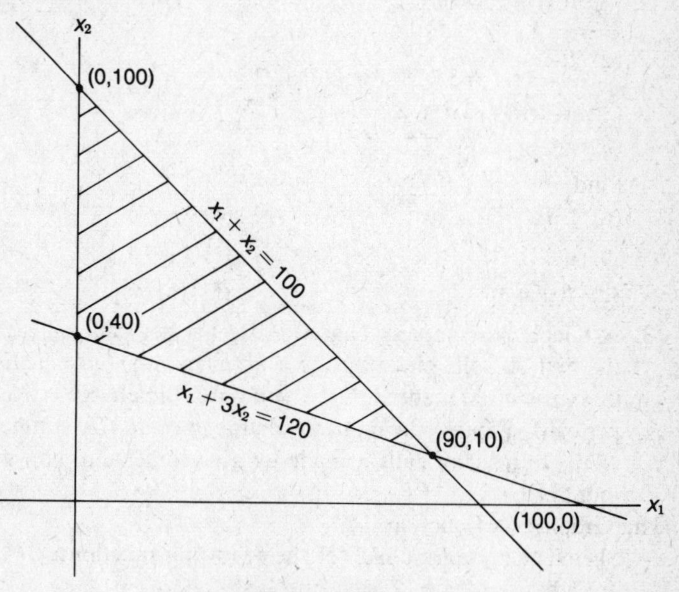

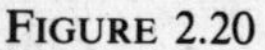
FIGURE 2.20

Now, by a well-known theorem, the maximum and minimum values of a linear function whose variables are subject to a set of linear inequalities occur at corner points. Hence the solution to our problem can be obtained by evaluating $f(x_1,x_2)$ at each of the three corner points and selecting the point that produces the largest value. Thus we have

$$f(0,100) = 30{,}000$$
$$f(0,40) = 12{,}000$$
$$f(90,10) = 39{,}000$$

Hence the maximum profit can be realized by manufacturing 90 *Scramblers* and 10 *Roadroyals* a week.

Problems—Chapter 2

Functions in General

1. Find the domain and the range of each of the following functions (use interval notation):
   (a) $f = \{(2,5), (3,7), (4,9)\}$
   (b) $f(x) = \sqrt{7 - x}$
   (c) $f(x) = -\sqrt{16 - x^2}$
   (d) $f(x) = \dfrac{1}{x+3}$
   (e) $f(x) = \dfrac{1}{(x-2)^2}$
   (f) $f(x) = \dfrac{1}{\sqrt{x^2 - 25}}$
2. Given $f(x) = 3x^2 + 2x - 4$
   $g(x) = \sqrt{x+4}$
   Find:
   (a) $f(0)$
   (b) $g(12)$
   (c) $f(-p)$
   (d) $g(-4)$
   (e) $f[g(x)]$
   (f) $g[f(x)]$
3. A telephone company charges 10 cents per call for each of the first 30 calls made during a given month. For each call in excess of 30 made during the month, the charge is 7 cents per call. Express the monthly charge in cents C as a function of the number of calls x made by a customer during a given month.

The Graph of a Function

4. Sketch the graph of each of the following functions:
   (a) $y = x^2 - 4$
   (b) $y = \sqrt{x^2 - 16}$
   (c) $y = \dfrac{1}{x+3}$
   (d) $y = |x^2 - 4|$

The Limit of a Function

5. Evaluate each of the following limits:
   (a) $\lim\limits_{x\to 2}\left(x - \dfrac{1}{x}\right)$
   (b) $\lim\limits_{x\to 0}\dfrac{x-2}{x+4}$
   (c) $\lim\limits_{x\to 2}\dfrac{x-2}{x+4}$
   (d) $\lim\limits_{x\to \infty}\left(3 - \dfrac{2}{x-1}\right)$
   (e) $\lim\limits_{x\to \infty}\dfrac{3x^2 - 4x + 5}{x^2 - 7}$
   (f) $\lim\limits_{x\to 4}\dfrac{x^2 - 16}{x-4}$

(g) $\lim_{x \to -3} \frac{x^2 - x - 12}{x + 3}$

(h) $\lim_{x \to \infty} \frac{4x - 5}{x^2 - 6}$

(i) $\lim_{x \to 2} \frac{x^2 + 3x - 10}{x^2 - 8x + 12}$

(j) $\lim_{x \to a} \frac{x^2 - ax}{x^2 - a^2}$

The Inverse of a Function

6. Determine which of the following functions have inverses and determine the inverse if it exists:
   (a) $f = \{(2,-4), (3,5), (0,-7)\}$
   (b) $f = \{(4,5), (6,8), (8,5)\}$
   (c) $f(x) = 2x - 7$
   (d) $f(x) = 4x^2 - 6$
   (e) $f(x) = (x - 6)^3$

Linear Functions

7. Write, in the form $Ax + By + C = 0$, the equations of the lines satisfying the following conditions:
   (a) Slope $= -3$, y intercept $= 4$
   (b) Passes through the points $(-2,6)$ and $(5,-2)$
   (c) Passes through the points $(4,5)$ and $(4,-3)$
   (d) x intercept $= 4$, y intercept $= -3$
8. Find the equation of the line that is parallel to the line $3x - 5y - 7 = 0$ and that contains the point $(2,-3)$. (Hint: first find the slope of the given line.)
9. Find the equation of the line that is perpendicular to the line $x - 3y + 4 = 0$ and that contains the point $(-5,4)$.

Distance and Point of Division Formulas

10. Find the distance between the following pairs of points:
    (a) $(5,7)$ and $(-3,7)$
    (b) $(2,5)$ and $(6,8)$
    (c) $(p,2q)$ and $(2p,4q)$
11. For the line segment connecting the points P_1 $(-2,5)$ and P_2 $(6,9)$, find:
    (a) The midpoint
    (b) The point one-quarter of the way from P_1 to P_2
    (c) The point P such that $\overline{P_1P}/\overline{PP_2} = 7/9$

The Quadratic Function and Its Graph

12. Sketch the graph of each of the following quadratic functions. First determine the highest or lowest point on each curve, and then plot several other points:
    (a) $y = 3x^2 - 12x + 5$
    (b) $y = x^2 + 8x - 10$
    (c) $y = -5x^2 + 10x - 4$
    (d) $y = -x^2 - x + 3$

The Logarithmic Function

13. Solve each of the following equations:
    (a) $\log_5 x = 3$
    (b) $\log_x\left(\frac{1}{16}\right) = 4$
    (c) $\log(x-1) + \log(x+4) = \log 6$
    (d) $\log_4 x - \log_4(x-3) = 2$
    (e) $4^x = 42$

The Conics

14. Find the center and the radius of each of the following circles:
    (a) $x^2 + 4x + y^2 - 14y + 37 = 0$
    (b) $x^2 + 10x + y^2 - 4y + 29 = 0$
    (c) $4x^2 + 4y^2 - 16x - 12y - 11 = 0$
15. For the following ellipse, find the center, the lengths of the major and minor axes, and the eccentricity:
    $$x^2 + 9y^2 - 2x + 54y + 46 = 0$$
16. Find the center and the eccentricity of the following hyperbola:
    $$16x^2 - 9y^2 + 160x + 54y + 175 = 0$$
17. Find the vertex of each of the following parabolas and state the direction in which each parabola opens:
    (a) $x^2 - 8x - 8y - 8 = 0$
    (b) $6x - y^2 - 10y - 37 = 0$
    (c) $x^2 + 10y + 70 = 0$
18. Solve each of the following systems of equations:
    (a) $x^2 + 2y^2 = 11$
    $3x + y = -8$
    (b) $x^2 + y^2 = 40$
    $y = \frac{3}{2}x^2$
    (c) $x^2 + 2y^2 = 34$
    $x^2 - y^2 = 7$
19. Sketch the graph of the following system of linear inequalities:
    $$4x + 3y \geq 0$$
    $$3x - y \geq 2$$
    $$x \leq 2$$
20. Find the maximum value and the minimum value of the function, $f(x,y) = 2x + y$, if x and y are subject to the system of inequalities specified in Problem 19.
21. A company employs workers for two types of jobs, A and B. Because of space limitations, the maximum number of

job A workers is 40, and the maximum number of job B workers is 30. Job A workers are paid \$3/h and job B workers are paid \$4/h. The company's budget permits a maximum payroll of \$192/h. If the company wishes to benefit the community in which it is located by employing the maximum total number of workers, how many should be employed for each job type?

3

TRIGONOMETRIC AND HYPERBOLIC FUNCTIONS

3.1 Definitions of the Trigonometric Functions

The simplest definitions of the trigonometric functions are those which relate to the unit circle, i.e., a circle with center at the origin of the rectangular coordinate system and with radius equal to 1. The equation of this circle is of course $x^2 + y^2 = 1$. See Figure 3.1.

Consider now an arc of length s extending from the point (1,0) to the point (x,y) as shown in the figure, i.e., an arc obtained by starting at the point (1,0) and proceeding in the counterclockwise direction. The length of such an arc is taken to be a positive number. If the arc had been obtained by starting at (1,0) and proceeding in the clockwise direction, its length would be taken as a negative number. Furthermore, since it is assumed that we may generate an arc of any length, either positive or negative, by proceeding around the circle in the appropriate direction, s may assume any real value whatsoever. For any arc s the initial point is (1,0) and the terminal point is a point on the circle whose coordinates are (x,y). We now define the following functions:

$$\text{cosine } s = x \tag{1}$$

$$\text{sine } s = y \tag{2}$$

In other words, the abscissa and the ordinate of the terminal point of s serve to define two functions of s that are called **cosine s** and **sine s,** respectively. Note that each of these satisfies the definition of a function, since corresponding to each value of s there will be one and only one value of the cosine and one and only one value of the sine.

We have already seen that the domain of each of these functions is the set of all real numbers. Since x and y are coordinates of

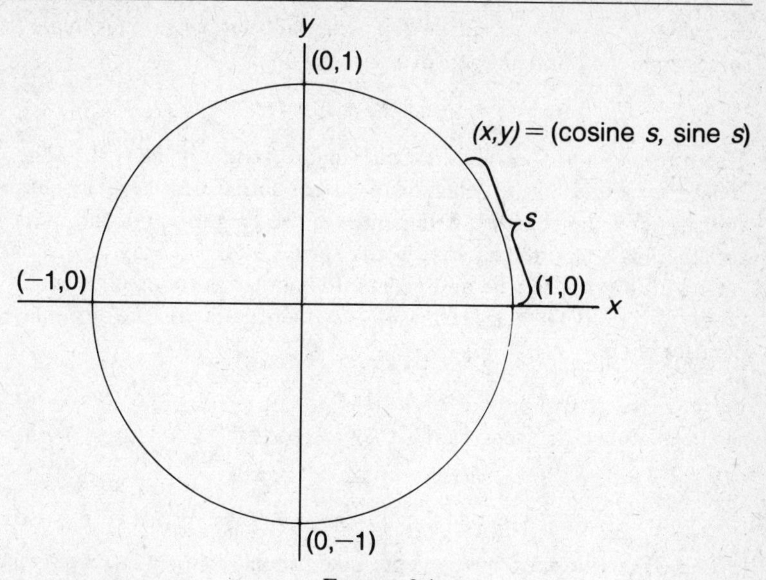

FIGURE 3.1

points on the unit circle, it is clear that both x and y lie in the closed interval $[-1,1]$. Hence this interval constitutes the range of both the cosine and sine functions.

Because of their close connection with the unit circle, the cosine and sine functions (as well as certain other functions to be defined presently) are sometimes referred to as the **circular functions,** although the designation **trigonometric** is more commonly used. It should be noted that although trigonometric functions are often associated with angles, the functions that we have just defined have no necessary connection with angles. They are simply functions whose domains consist of real numbers. In Section 3.5 we consider an alternative way of defining the trigonometric functions.

The sine and cosine functions are employed extensively in applications of mathematics to certain problems in mechanics and electricity. This is due to the fact that they have a repetitive, or **periodic,** property, which will be discussed in the next section.

An example of a mechanical problem whose solution involves the sine and cosine functions concerns a mass suspended from a spring. The mass is displaced from its equilibrium position and released either from rest or with an initial velocity. It can be shown that if there are no frictional or other resisting forces acting on the system, the subsequent displacement y from the equilibrium position is related to the time t by an equation of the form

$$y = A \text{ cosine } Kt + B \text{ sine } Kt$$

The motion in this case is called **simple harmonic motion.** The constants A and B are determined by the initial displacement and velocity, and the constant K depends on the magnitude of the mass and the force required to stretch the spring a unit of length.

In addition to the sine and cosine, four other trigonometric functions may be defined in terms of the coordinates of the terminal point of the arc s.

$$\text{tangent } s = y/x \qquad x \neq 0 \tag{3}$$

$$\text{cosecant } s = 1/y \qquad y \neq 0 \tag{4}$$

$$\text{secant } s = 1/x \qquad x \neq 0 \tag{5}$$

$$\text{cotangent } s = x/y \qquad y \neq 0 \tag{6}$$

These functions also have the periodic property previously referred to in connection with the sine and cosine functions. However, while the sine and cosine functions are continuous for all values of s, the four functions defined above are discontinuous because they are undefined for certain values of s.

The names of the six trigonometric functions are frequently abbreviated as follows:

$$\text{sine } s = \sin s \qquad \text{cosecant } s = \csc s$$

$$\text{cosine } s = \cos s \qquad \text{secant } s = \sec s$$

$$\text{tangent } s = \tan s \qquad \text{cotangent } s = \cot s$$

3.2 Properties of the Trigonometric Functions

We shall now discuss the periodic property of the sine and cosine functions. In general, a function $f(x)$ is said to be periodic of period p if there exists a positive number p such that $f(x+p) = f(x)$ and if p is the smallest positive number for which this relationship holds.

The circumference C of a circle of radius r is given by the formula $C = 2\pi r$. Therefore, the circumference of the unit circle is 2π. Hence, if s_1 and s_2 are two arcs as defined above such that $s_2 = s_1 + 2\pi$, then s_1 and s_2 have the same terminal point. Therefore, in view of the manner in which we have defined the sine and cosine functions, we have, for any s,

$$\sin(s + 2\pi) = \sin s$$
$$\cos(s + 2\pi) = \cos s$$

Hence the sine and cosine functions are periodic with period 2π. From the manner in which the cosecant and secant functions have been defined [Equations (4) and (5)] it is apparent that they are the reciprocals of the sine and cosine, respectively, and hence they are also periodic with period 2π. It can be shown that the tangent and cotangent functions are periodic with period π.

Another important property of functions in general that is possessed by the trigonometric functions is that of being either even or odd. A function $f(x)$ is said to be an **even function** if for all x in its domain $f(-x) = f(x)$. A function $f(x)$ is said to be an **odd function** if for all x in its domain $f(-x) = -f(x)$.

For example, the function $f(x) = 3x^4 - 2x^2$ is an even function since $f(-x) = 3(-x)^4 - 2(-x)^2 = 3x^4 - 2x^2 = f(x)$. On the other hand, the function $g(x) = 4x^3 + x$ is an odd function since $g(-x) = 4(-x)^3 + (-x) = -4x^3 - x = -g(x)$. It may be noted that many functions are neither even nor odd. One example is the function $h(x) = x + 2$, since $h(-x) = -x + 2$, which is neither $h(x)$ nor $-h(x)$.

We shall now show that the sine function is odd and that the cosine function is even. In Figure 3.2, two arcs s_1 and s_2 that are equal in length but opposite in direction have been indicated. Clearly $s_2 = -s_1$. Also from the symmetry of the circle, it is apparent that $x_2 = x_1$ and $y_2 = -y_1$. Therefore, $\sin(-s_1) = \sin s_2 = y_2 = -y_1 = -\sin s_1$. Also $\cos(-s_1) = \cos s_2 = x_2 = x_1 = \cos s_1$. Hence the assertion made in the preceding paragraph has been proved.

It follows immediately that the cosecant function is odd and that the secant function is even. Furthermore $\tan(-s_1) = \tan s_2 = y_2/x_2 = -y_1/x_1 = -\tan s_1$. Hence, the tangent and cotangent functions are odd.

Another interesting property of one of the trigonometric functions that can be readily proved using the unit circle definition is the following:

$$\cos(s_1 + s_2) = \cos s_1 \cos s_2 - \sin s_1 \sin s_2 \qquad (1)$$

The proof may be obtained by selecting appropriate points on the unit circle and employing the distance formula (Section 2.10). As will be indicated later, formula (1) leads to a host of additional

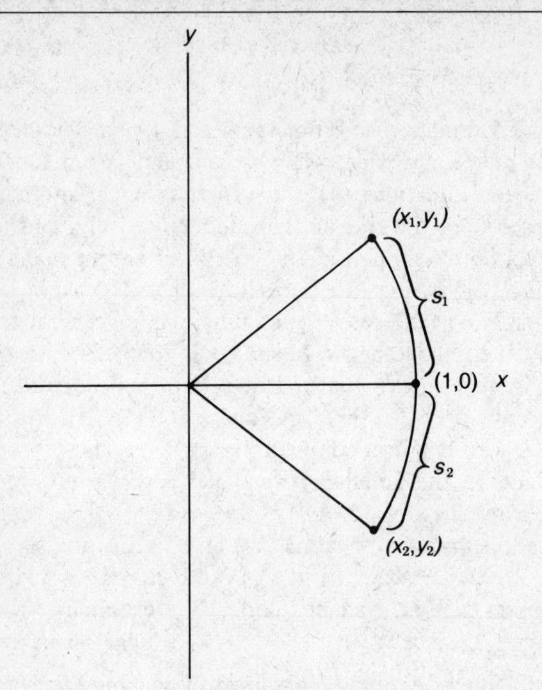

FIGURE 3.2

important trigonometric relationships. For example, using the fact that the cosine function is even and the sine function is odd, we obtain:

$$\begin{aligned}\cos(s_1 - s_2) &= \cos[s_1 + (-s_2)] \\ &= \cos s_1 \cos(-s_2) - \sin s_1 \sin(-s_2) \\ &= \cos s_1 \cos s_2 + \sin s_1 \sin s_2 \qquad (2)\end{aligned}$$

3.3 Some Values of the Trigonometric Functions

A number of values of the trigonometric functions corresponding to specified values of s can be readily obtained with reference to the unit circle as shown in Figure 3.1. For example, by referring to the four points at which the circle intersects the coordinate axes we obtain a number of values. The point (1,0) corresponds to an arc s of length 0. Hence we conclude that $\sin 0 = 0$, $\cos 0 = 1$, $\tan 0 = 0/1 = 0$, $\sec 0 = 1/1 = 1$. It is clear that the cosecant and cotangent functions are undefined at $s = 0$ since their definitions in this case would each involve a fraction with a 0 in the denominator.

The point (0,1), since it is reached by proceeding one-quarter of the way around the circle from the initial point (1,0), corresponds to $s = 2\pi/4 = \pi/2$. Hence, we observe that $\sin \pi/2 = 1$, $\cos \pi/2 = 0$, $\csc \pi/2 = 1$, $\cot \pi/2 = 0$, and that the secant and tangent functions are undefined at this point.

Using similar reasoning, we readily reach the conclusion that $\sin \pi = 0$, $\cos \pi = -1$, $\tan \pi = 0$, $\sec \pi = -1$, and that the cosecant and cotangent functions are undefined for $s = \pi$. Also we readily ascertain that $\sin 3\pi/2 = -1$, $\cos 3\pi/2 = 0$, $\cot 3\pi/2 = 0$, $\csc 3\pi/2 = -1$, and that the tangent and secant functions are undefined for $s = 3\pi/2$.

Using the periodic properties of the trigonometric functions, we can obtain an unlimited number of additional values from those already listed. For example, $\sin 5\pi/2 = \sin (\pi/2 + 2\pi) = \sin \pi/2 = 1$. Also $\cos 3\pi = \cos (\pi + 2\pi) = \cos \pi = -1$.

We may also make use of the even and odd properties of the trigonometric functions to obtain such values as $\sin (-3\pi/2) = -\sin 3\pi/2 = -(-1) = 1$ and $\cos (-\pi/2) = \cos \pi/2 = 0$.

Of course the value of any trigonometric function corresponding to a given value of s could be obtained by making an accurate measurement of s and the abscissa and ordinate of its terminal point. However this process would be tedious and impractical. In practice, the calculation of the trigonometric functions is carried out by a method to be explained briefly in a later section.

3.4 Angles and Their Measure

It has already been pointed out that the domain of the trigonometric functions consists of the set of all real numbers and that these numbers have no necessary connection with angles. However, for certain purposes, it is useful to think of these numbers as a certain type of angle measure. This leads us into a discussion of angles and methods of measuring them.

An **angle** may be defined as a measure of the amount of rotation of a line segment about a point. An angle generated by a counterclockwise rotation is taken as positive and one generated by a clockwise rotation as negative. The rotation can take place to any extent in either direction. Three angles are represented in Figure 3.3.

In each case the arrowhead indicates the direction of the rotation, so that the angles in (a) and (b) are positive, while the angle in (c)

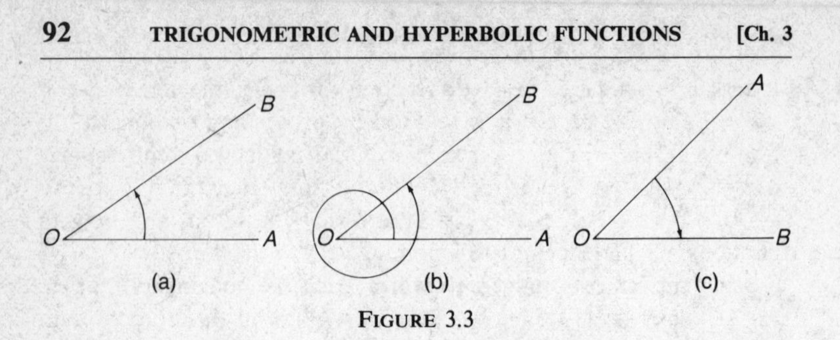

FIGURE 3.3

is negative. Note also that the angle represented by (b) involves more than a complete rotation. In each case the side $\overline{OA}$ is called the initial side of the angle, the side $\overline{OB}$ is called the terminal side, and the point O is called the vertex.

There are two units of measurement for angles in common use, the degree and the radian. A degree is defined as 1/360 part of a complete rotation. Note that the use of the number 360 in this definition is completely arbitrary. An angle representing one-half of a complete rotation, i.e., 180° is commonly called a **straight angle,** and an angle representing one-fourth of a complete rotation, i.e., 90°, is called a **right angle.**

The **radian** measure of an angle is defined with the vertex of the angle coinciding with the center of a circle. If the arc intercepted by the sides of the angle on the circumference of the circle is designated by s and the radius of the circle by r, then the radian measure of the angle is defined as the ratio of s to r. This ratio is easily seen to be independent of the size of the circle. Hence, if the radian measure of the angle is designated by θ, we have

$$\theta = \frac{s}{r}$$

See Figure 3.4.

Although the degree measure of angles is much better known to people in general than the radian measure, the radian measure is more important in mathematics. This might be intuitively expected since the radian measure is free of the arbitrary element referred to in the previous discussion of the degree measure. Radian measure is used almost exclusively in the calculus of the trigonometric functions; the reason for this will be made clear later on.

However, since both degree and radian measures are used, it is sometimes necessary to convert the degree measure of an angle to its radian measure or vice-versa. This is easily accomplished as

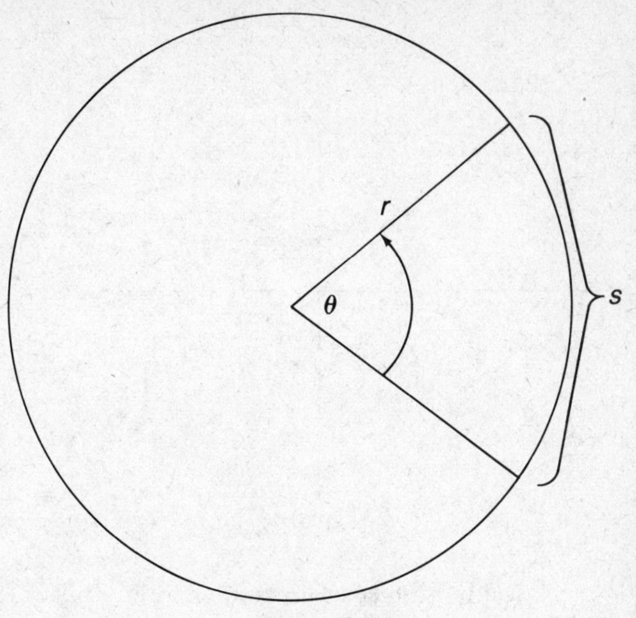

FIGURE 3.4

follows. The radian measure of an angle is simply the number of times the radius of the circle is contained in the intercepted arc. Since the radius is contained in the entire circumference of the circle 2π times and since this corresponds to a complete rotation, we obtain the relationship

$$360° = 2\pi \text{ radians}$$

From this relationship the radian equivalent of many commonly used angles can easily be obtained. For example, $180° = \pi$ radians, $90° = \pi/2$ radians, $45° = \pi/4$ radians, etc. Furthermore, 1 radian $= 180°/\pi \simeq 57.3°$ and $1° = \pi/180 \simeq 0.017$ radians.

Since radian measure is so widely used in mathematics, it is common practice that if no measure of an angle is specified, then radian measure is understood. Thus $\pi/6$ is understood to refer to an angle of $\pi/6$ radians, or 30°.

3.5 Alternative Definitions of the Trigonometric Functions

We now make use of the rectangular coordinate system to establish some useful alternative definitions of the trigonometric functions. To do this we first define the concept of an angle in standard position.

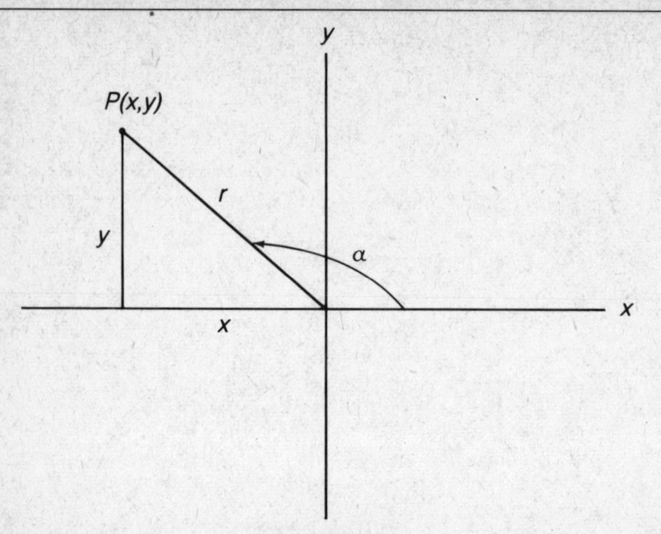

FIGURE 3.5

An angle is said to be in **standard position** with reference to the rectangular coordinate system if its vertex is located at the origin and if its initial side coincides with the positive half of the x axis. An angle α in standard position is shown in Figure 3.5

When placed in standard position, an angle is said to be in the quadrant in which its terminal side lies. Two angles with the same terminal side are said to be **coterminal.** The angle α shown in Figure 3.5 is therefore said to be in quadrant II. Angles whose terminal sides coincide with the coordinate axes are referred to as the **quadrantal angles.**

We now choose any point P except the origin on the terminal side of α, and we associate with α three numbers: x, the abscissa of P; y, the ordinate of P; and r, the distance from the origin to P. Note that r is always considered to be a positive number.

We now define the six trigonometric functions of α as follows:

$$\sin \alpha = \frac{y}{r} \qquad \csc \alpha = \frac{r}{y}$$

$$\cos \alpha = \frac{x}{r} \qquad \sec \alpha = \frac{r}{x}$$

$$\tan \alpha = \frac{y}{x} \qquad \cot \alpha = \frac{x}{y}$$

It is easily shown, using similar triangles, that each of the six ratios shown above is independent of the particular point P that is selected on the terminal side of α. Furthermore, the trigonometric functions of any two coterminal angles will evidently be the same.

If these definitions of the trigonometric function are compared with those given in Section 3.1, it can readily be seen that the functions defined previously are identical to those defined above. In view of the definition of the radian measure of an angle that has been given, it is readily seen that the arc s used in the unit circle definitions of the trigonometric functions is simply the radian measure of an angle in standard position, since if $r = 1$, the ratio $s/r = s$. Furthermore, since in connection with the definitions given in this section, the point P on the terminal side of the angle in standard position can be any point except the origin, P may be chosen at unit length from the origin. Then, $y/r = y$ and $x/r = x$, where x and y are the coordinates of P. Thus the two definitions of the functions that have been given are equivalent, and the value of s used in connection with unit circle definitions can be thought of either as the radian measure of an angle or simply as a real number.

For acute angles, i.e., angles between 0° and 90°, there is a well-known set of definitions of the trigonometric functions associated

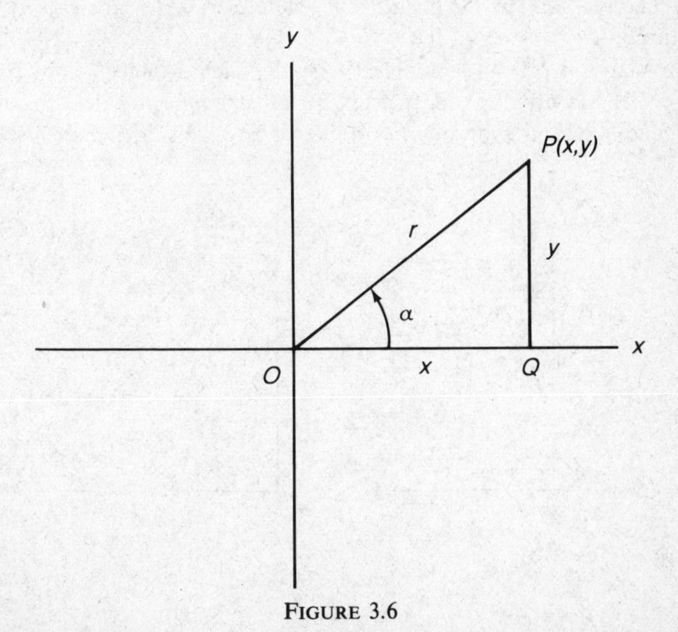

FIGURE 3.6

with the sides of a right triangle. These can be shown by drawing an angle in standard position in the first quadrant as shown in Figure 3.6.

The triangle OQP is a right triangle, and with reference to the angle α, the side OQ is referred to as the side **adjacent** to α, and the side PQ as the side **opposite** to α. The side OP, which is opposite the right angle, is called the **hypotenuse.**

Evidently, using the general definitions set forth above, the trigonometric functions of α can now be stated as follows:

$$\sin \alpha = \frac{\text{opposite side}}{\text{hypotenuse}} \qquad \csc \alpha = \frac{\text{hypotenuse}}{\text{opposite side}}$$

$$\cos \alpha = \frac{\text{adjacent side}}{\text{hypotenuse}} \qquad \sec \alpha = \frac{\text{hypotenuse}}{\text{adjacent side}}$$

$$\tan \alpha = \frac{\text{opposite side}}{\text{adjacent side}} \qquad \cot \alpha = \frac{\text{adjacent side}}{\text{opposite side}}$$

3.6 Special Values of the Trigonometric Functions

The right-triangle definitions of the trigonometric functions may be used in connection with two reference triangles to obtain all of the functions of 30°, 45°, and 60°. These triangles are shown in Figure 3.7.

In Figure 3.7(a) an equilateral triangle, each of whose sides is of length 2, has been divided into two right triangles by drawing the perpendicular bisector of one of its sides. This line also bisects

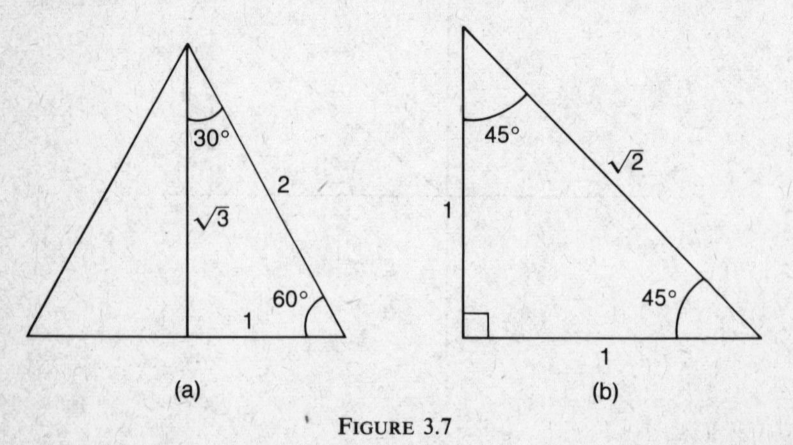

FIGURE 3.7

the angle from whose vertex it has been drawn, and its length has been obtained by the Pythagorean theorem. All of the functions of 30° and 60° can now be readily obtained by referring to this figure. Figure 3.7(b) represents an isosceles right triangle whose equal sides are each of length 1. By the Pythagorean theorem its hypotenuse is found to be $\sqrt{2}$. Each of the functions of 45° can be obtained with reference to this triangle.

If all the values of the trigonometric functions that were obtained from the unit circle definitions are combined with those that can be obtained from the two reference triangles just presented, we may construct the following table.

Radian measure	0	$\pi/6$	$\pi/4$	$\pi/3$	$\pi/2$	π	$3\pi/2$
Degree measure	0°	30°	45°	60°	90°	180°	270°
sin	0	1/2	$1/\sqrt{2}$	$\sqrt{3}/2$	1	0	−1
cos	1	$\sqrt{3}/2$	$1/\sqrt{2}$	1/2	0	−1	0
tan	0	$1/\sqrt{3}$	1	$\sqrt{3}$	undef.	0	undef.
csc	undef.	2	$\sqrt{2}$	$2/\sqrt{3}$	1	undef.	−1
sec	1	$2/\sqrt{3}$	$\sqrt{2}$	2	undef.	−1	undef.
cot	undef.	$\sqrt{3}$	1	$1/\sqrt{3}$	0	undef.	0

The angles listed in the above table are often called the **special angles** since they occur quite frequently in applications of trigonometry, and the student should be thoroughly familiar with all of the values shown in the table.

3.7 The Reference Angle

A concept of importance in the determination of the values of the functions of many angles is that of the reference angle. The **reference angle** associated with a given angle α in standard position is defined as the acute angle between the terminal side of α and the x axis. The reference angle β is shown in Figure 3.8 for several angles α in quadrants II, III, and IV, where $0 \leq \alpha < 360°$.

From these diagrams it is readily seen that the following relationships hold between α and its reference angle β, depending on the quadrant in which α lies.

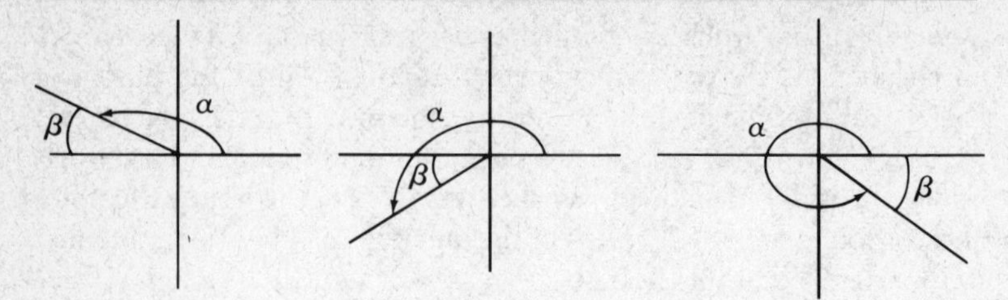

FIGURE 3.8

QUADRANT	II	III	IV
Degree measure	$\beta = 180° - \alpha$	$\beta = \alpha - 180°$	$\beta = 360° - \alpha$
Radian measure	$\beta = \pi - \alpha$	$\beta = \alpha - \pi$	$\beta = 2\pi - \alpha$

For example, the reference angles associated with 150°, 225°, and 300° are 30°, 45°, and 60°, respectively. The same statement may be made using radian measure: the reference angles associated with $5\pi/6$, $5\pi/4$, and $5\pi/3$, are $\pi/6$, $\pi/4$, and $\pi/3$, respectively. The reference angle for an angle outside the range $0 \leq \alpha < 360°$ can be obtained by first adding or subtracting an integral multiple of 360° to obtain a coterminal angle in the specified range. For example, the reference angle for 840° is the same as the reference angle for $840° - 720° = 120°$.

The importance of the reference angle lies in the fact stated in the following theorem.

Theorem I. Any function of a given angle is numerically equal, i.e., equal in absolute value, to the same function of its reference angle.

We shall prove the theorem for an angle lying in quadrant II, and the proof can readily be generalized. Figure 3.9 shows an angle α in standard position, along with its reference angle β.
Using the general definitions of the trigonometric functions for α and the right-triangle definitions for β, we see that the functions of α and corresponding functions of β will be numerically equal. For example,

$$\sin \alpha = \frac{y}{r} = \frac{\overline{PQ}}{\overline{OP}} = \sin \beta$$

$$\cos\alpha = \frac{x}{r} = -\frac{\overline{OQ}}{\overline{OP}} = -\cos\beta$$

$$\tan\alpha = \frac{y}{x} = -\frac{\overline{PQ}}{\overline{OQ}} = -\tan\beta \qquad \text{etc.}$$

The fact stated in Theorem I enables us to calculate readily the values of the trigonometric functions of many angles provided that we know the values of the same function of the reference angles. Since the theorem provides us with the numerical value of the function, we need only provide the proper sign to specify the value of the function.

The signs of the functions of angles in the various quadrants can be readily determined by noting the signs of the quantities used to define these functions. We note that r is always positive, that in quadrant I, x and y are both positive, in quadrant II, x is negative and y is positive, in quadrant III, x and y are both negative, and in quadrant IV, x is positive and y is negative. Using these facts along with the definitions of the functions, we conclude that in quadrant I, all of the functions are positive, that the sin and csc are positive in quadrant II, the tan and cot are positive in quadrant III, and the cos and sec are positive in quadrant IV. All of the other functions are negative in the specified quadrants. These facts are summarized in Figure 3.10.

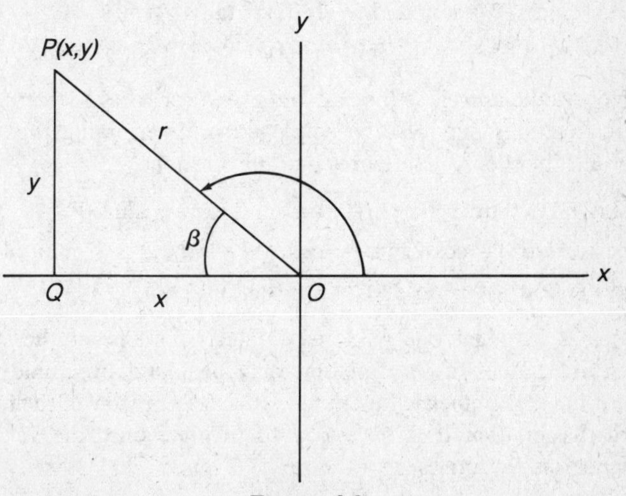

FIGURE 3.9

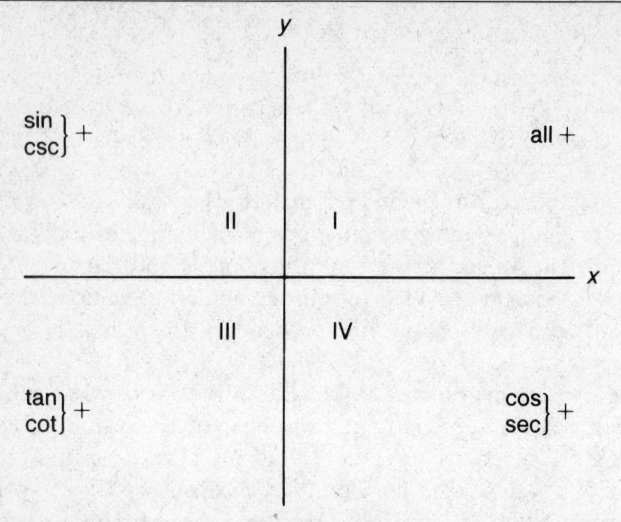

FIGURE 3.10

Here are some examples of functional values obtained by the process outlined above:

$$\sin 150° = \sin (180° - 150°) = \sin 30° = \tfrac{1}{2}$$
$$\cos 120° = -\cos 60° = -\tfrac{1}{2}$$
$$\tan 225° = \tan (225° - 180°) = \tan 45° = 1$$
$$\tan 300° = -\tan (360° - 300°) = -\tan 60° = -\sqrt{3}$$

The periodic as well as the odd and even properties of the trigonometric functions may also be employed in determining the values of certain functions. Here are some illustrations.

$$\sin 960° = \sin (960° - 720°) = \sin 240° = -\sin 60° = -\sqrt{\tfrac{3}{2}}$$
$$\cos (-225°) = \cos 225° = -\cos 45° = -1\sqrt{2}$$
$$\tan (-330°) = -\tan 330° = -(-\tan 30°) = 1/\sqrt{3}$$

In cases in which the reference angle is not one of the special angles, the values of the function may be found by employing a table of the trigonometric functions. It is clear in view of the foregoing discussion that it is necessary to tabulate only the values of the functions for angles α in the range $0 \leq \alpha \leq 90°$.

3.8 Trigonometric Identities

It will be recalled that in connection with algebraic functions an algebraic identity was defined as an equation for which the solution set consists of the set of all numbers that produce a real number when substituted in either side of the equation. Trigonometric identities are similarly defined.

There are eight trigonometric identities that are often referred to as the fundamental identities. The first three of these are a direct consequence of the definitions of the trigonometric functions:

$$\csc x = \frac{1}{\sin x} \tag{1}$$

$$\sec x = \frac{1}{\cos x} \tag{2}$$

$$\tan x = \frac{1}{\cot x} \tag{3}$$

The next two may be easily obtained as follows:

$$\tan x = \frac{y}{x} = \frac{y}{r} \div \frac{x}{r} = \frac{\sin x}{\cos x} \tag{4}$$

$$\cot x = \frac{x}{y} = \frac{x}{r} \div \frac{y}{r} = \frac{\cos x}{\sin x} \tag{5}$$

The remaining three are often called the Pythogorean identities since they follow directly from the Pythagorean theorem, as shown below.

Referring to Figure 3.5, which shows an angle in standard position, we note that the quantities x, y, and r are related by the equation $x^2 + y^2 = r^2$. Hence we have

$$\sin^2 x + \cos^2 x = \frac{x^2}{r^2} + \frac{y^2}{r^2} = \frac{x^2 + y^2}{r^2} = \frac{r^2}{r^2} = 1 \tag{6}$$

If we divide each term of the first and last members of Equation (6) by $\cos^2 x$, we obtain

$$\frac{\sin^2 x}{\cos^2 x} + \frac{\cos^2 x}{\cos^2 x} = \frac{1}{\cos^2 x}$$

or

$$\tan^2 x + 1 = \sec^2 x \tag{7}$$

Finally, if each term of (6) is divided by $\sin^2 x$, we obtain

$$\frac{\sin^2 x}{\sin^2 x} + \frac{\cos^2 x}{\sin^2 x} = \frac{1}{\sin^2 x}$$

or

$$1 + \cot^2 x = \csc^2 x \tag{8}$$

It should be noted that expressions such as $\sin^2 x$ stand for the square of $\sin x$, i.e., for $(\sin x)^2$.

The eight fundamental identities that have been listed may be used to prove innumerable other identities or to simplify certain trigonometric expressions. A few examples will now be furnished.

Example 1 Simplify

$$\frac{1}{1 + \sin \alpha} + \frac{1}{1 - \sin \alpha}$$

$$\frac{1}{1 + \sin \alpha} + \frac{1}{1 - \sin \alpha} = \frac{1 - \sin \alpha + 1 - \sin \alpha}{1 - \sin^2 \alpha} = \frac{2}{1 - \sin^2 \alpha}$$

$$= \frac{2}{\cos^2 \alpha} = 2 \sec^2 \alpha$$

The replacement of $1 - \sin^2 \alpha$ by $\cos^2 \alpha$ follows, of course, from Equation (6).

Example 2 Simplify

$$\sin^4 \alpha - \cos^4 \alpha$$

$$\sin^4 \alpha - \cos^4 \alpha = (\sin^2 \alpha + \cos^2 \alpha)(\sin^2 \alpha - \cos^2 \alpha)$$
$$= \sin^2 \alpha - \cos^2 \alpha = \sin^2 \alpha - (1 - \sin^2 \alpha) = 2 \sin^2 \alpha - 1$$

Example 3 Simplify

$$\frac{1 + \tan^2 \alpha}{\csc^2 \alpha}$$

$$\frac{1 + \tan^2 \alpha}{\csc^2 \alpha} = \frac{\sec^2 \alpha}{\csc^2 \alpha} = \frac{1/\cos^2 \alpha}{1/\sin^2 \alpha} = \frac{\sin^2 \alpha}{\cos^2 \alpha} = \tan^2 \alpha$$

Example 4 Simplify

$$\tan^2 \alpha \csc^2 \alpha - \tan^2 \alpha$$

$$\tan^2 \alpha \csc^2 \alpha - \tan^2 \alpha = \tan^2 \alpha(\csc^2 \alpha - 1) = \tan^2 \alpha \cot^2 \alpha = 1$$

3.9 Trigonometric Equations

Although in most cases the solution sets of trigonometric equations are infinite, we shall restrict these sets to angles in the range $0° \leq$

$\alpha < 360°$ (or in radians, $0 \leq x < 2\pi$). These angles are often called the **primary angles.**

Consider the equation $2 \sin \alpha - 1 = 0$. Evidently this is equivalent to the equation $\sin \alpha = \frac{1}{2}$. Since α is an angle whose sine is positive, it evidently lies in either quadrant I or II. The quadrant I angle whose sine is $\frac{1}{2}$ is 30°. Recalling the formula for the reference angle for an angle in quadrant II, $\beta = 180° - \alpha$, we see that $\alpha = 180° - \beta = 180° - 30° = 150°$. Hence the solution set (confined to primary angles) consists of the values 30° and 150°, or in radian measure $\pi/6$ or $5\pi/6$.

Consider next the equation $2 \cos^2 \alpha - 3 \cos \alpha - 2 = 0$. Factoring the left-hand side of this equation, we obtain $(2 \cos \alpha + 1)(\cos \alpha - 2) = 0$. Equating each factor to 0, we have

$$2 \cos \alpha + 1 = 0 \qquad \cos \alpha - 2 = 0$$
$$\cos \alpha = -\tfrac{1}{2} \qquad \cos \alpha = 2$$

Considering first the equation $\cos \alpha = -\frac{1}{2}$, we note that since α is an angle whose cosine is negative, α must lie in either quadrant II or III. The angle in quadrant I, whose cosine is $\frac{1}{2}$, is 60°. Hence this is the reference angle, and the solution set for this equation will include the values of α in quadrant II and quadrant III for which the reference angle is 60°. Using the same type of reasoning we applied in the previous example, we calculate these values of α to be 120° and 240°. Considering the other equation, $\cos \alpha = 2$, we see that the solution set to this equation is empty since the number 2 is outside the range of the cosine function. Hence the values 120° and 240° constitute the entire solution set of the original equation.

Sometimes we may employ the trigonometric identities in solving equations. Consider the equation $\sec^2 \alpha - 2 \tan \alpha = 0$. Using the identity $1 + \tan^2 \alpha = \sec^2 \alpha$, we obtain $\tan^2 \alpha - 2 \tan \alpha + 1 = 0$. This is equivalent to $(\tan \alpha - 1)^2 = 0$, or $\tan \alpha = 1$. The solution set evidently consists of angles in quadrants I and III and using the reference angle, 45°, we see that this set consists of the angles 45° and 225°.

If the solution of a trigonometric equation involves extracting the square root of both sides, care must be taken to include both the positive and the negative square roots. Consider, for example, the equation $4 \sin^2 \alpha - 3 = 0$. This is equivalent to $4 \sin^2 \alpha = 3$, or $\sin^2 \alpha = \frac{3}{4}$. Taking the square root of both sides, we obtain

$\sin \alpha = \pm\sqrt{3}/2$. Evidently in this case the solution sets consist of angles in all four quadrants. Since the reference angle is 60°, the solution set consists of the angles 60°, 120°, 240°, and 300°.

It will be noted that all of the examples we have considered could be solved using only the table of values of the special angles that was given in Section 3.6.

3.10 Additional Trigonometric Formulas

In connection with the unit circle definitions of the trigonometric functions we noted the important formula

$$\cos (s_1 + s_2) = \cos s_1 \cos s_2 - \sin s_1 \sin s_2$$

and we readily obtained from this the formula

$$\cos (s_1 - s_2) = \cos s_1 \cos s_2 + \sin s_1 \sin s_2$$

The arc lengths s_1 and s_2 used in these formulas can of course be replaced by any angles α and β.

Using the formula for $\cos (s_1 - s_2)$, we may readily obtain

$$\cos\left(\frac{\pi}{2} - \alpha\right) = \cos\frac{\pi}{2}\cos \alpha + \sin\frac{\pi}{2}\sin \alpha$$
$$= 0 \cdot \cos \alpha + 1 \cdot \sin \alpha = \sin \alpha \quad (1)$$

If we replace α by $\left(\frac{\pi}{2} - \alpha\right)$ in (1), we have

$$\sin\left(\frac{\pi}{2} - \alpha\right) = \cos \alpha \quad (2)$$

Then formulas (1) and (2) can be used to obtain

$$\sin (\alpha + \beta) = \sin \alpha \cos \beta + \cos \alpha \sin \beta \quad (3)$$

Replacing β by $-\beta$ in (3) leads to

$$\sin (\alpha - \beta) = \sin \alpha \cos \beta - \cos \alpha \sin \beta \quad (4)$$

Then the formulas for $\sin (\alpha + \beta)$ and $\cos (\alpha + \beta)$ may be used along with the identity $\tan \alpha = \sin \alpha/\cos \alpha$ to derive the formula

$$\tan (\alpha + \beta) = \frac{\tan \alpha + \tan \beta}{1 - \tan \alpha \tan \beta} \quad (5)$$

and, replacing β by $-\beta$, we obtain

$$\tan(\alpha - \beta) = \frac{\tan\alpha - \tan\beta}{1 + \tan\alpha\tan\beta} \tag{6}$$

If we replace β by π in (5), we obtain

$$\tan(\alpha + \pi) = \frac{\tan\alpha + \tan\pi}{1 - \tan\alpha\tan\pi} = \frac{\tan\alpha - 0}{1 - (\tan\alpha)(0)} = \tan\alpha \tag{7}$$

This equation partially substantiates the fact that the period of the tangent function is π.

The formulas presented thus far are generally referred to as the addition formulas. These may be used to establish what may be called the double-angle formulas. For example, the substitution of α for β in (3) leads to

$$\begin{aligned}\sin 2\alpha &= \sin(\alpha + \alpha) = \sin\alpha\cos\alpha + \cos\alpha\sin\alpha \\ &= 2\sin\alpha\cos\alpha\end{aligned} \tag{8}$$

Using the same procedure in connection with the formula for cos $(\alpha + \beta)$, we obtain

$$\begin{aligned}\cos 2\alpha &= \cos\alpha\cos\alpha - \sin\alpha\sin\alpha \\ &= \cos^2\alpha - \sin^2\alpha\end{aligned} \tag{9}$$

Two additional forms for the last formula may be obtained by using the identity $\sin^2\alpha + \cos^2\alpha = 1$, namely,

$$\cos 2\alpha = 1 - 2\sin^2\alpha \tag{10}$$

and

$$\cos 2\alpha = 2\cos^2\alpha - 1 \tag{11}$$

We can then use (11) to generate what may be called the half-angle formulas. Solving for $\cos^2\alpha$, we have $\cos^2\alpha = (1 + \cos 2\alpha)/2$. If we replace α by $\beta/2$ and take the square root of both sides of the equation, we obtain

$$\cos\frac{\beta}{2} = \pm\sqrt{\frac{1 + \cos\beta}{2}} \tag{12}$$

The selection of the positive or the negative sign on the right-hand side of (12) will depend on the quadrant in which $\beta/2$ lies.

In the same way we may use (10) to obtain

$$\sin\frac{\beta}{2} = \pm\sqrt{\frac{1 - \cos\beta}{2}} \tag{13}$$

Then, using (12) and (13), we have

$$\tan\frac{\beta}{2}=\frac{\sin\beta/2}{\cos\beta/2}=\pm\sqrt{\frac{1-\cos\beta}{1+\cos\beta}} \tag{14}$$

Two alternative forms for $\tan\beta/2$ are

$$\tan\frac{\beta}{2}=\frac{1-\cos\beta}{\sin\beta} \tag{15}$$

and

$$\tan\frac{\beta}{2}=\frac{\sin\beta}{1+\cos\beta} \tag{16}$$

We have thus seen how the formula for $\cos(s_1+s_2)$ leads to a series of formulas that are of considerable importance in the application of the trigonometric functions to other branches of mathematics.

Before closing this section on trigonometric formulas, we mention two additional identities that are of practical importance in the solution of triangles, i.e., in finding unknown sides and angles of triangles for which certain angles or sides are given. These formulas may be stated with reference to the triangle shown in Figure 3.11.

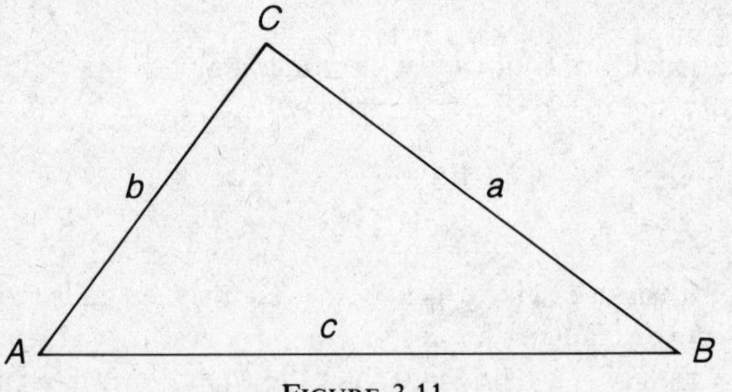

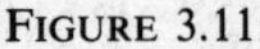

FIGURE 3.11

In this figure, which represents any triangle, the angles are designated by capital letters A, B, and C and the corresponding opposite sides by the small letters a, b, and c.

The unknown parts of any triangle for which sufficient information is given may be obtained using the law of sines or the law of cosines. The law of sines may be stated as follows:

$$\frac{a}{\sin A}=\frac{b}{\sin B}=\frac{c}{\sin C} \tag{17}$$

The law of cosines may be written

$$a^2 = b^2 + c^2 - 2bc \cos A \tag{18}$$

Evidently, (18) could be written in two additional forms. Although there are two additional laws that can be used in solving triangles, all cases can be handled with one of the two laws stated above. If the triangle involved happens to be a right triangle, then the laws of sines and cosines assume somewhat simpler forms. Note in particular that if the angle A is a right angle, formula (18) reduces to the Pythagorean thorem.

3.11 Tables and Graphs of the Trigonometric Functions

Tables of the values of the trigonometric functions at convenient intervals in the range from 0° to 90° are found in all standard trigonometry texts. It has already been shown that the use of the reference angle and the periodic properties of the trigonometric functions make it unnecessary to tabulate values for angles outside this range. Furthermore, the number of entries needed in the table can be halved in view of what may be called the complementary angle relationship of the functions of acute angles.

The sum of the two acute angles of any right triangle is 90°, and two such angles are said to be **complementary.** If we refer back to the right-triangle definitions of the trigonometric functions and if the acute angles are denoted by A and B so that $B = 90° - A$, then it is easily seen that the following relationships hold:

$$\sin A = \cos B = \cos (90° - A) \tag{1}$$

$$\sec A = \csc B = \csc (90° - A) \tag{2}$$

$$\tan A = \cot B = \cot (90° - A) \tag{3}$$

Hence the entry in the table for sin 20° also serves for cos 70°; the entry for tan 50° also serves for cot 40°, etc.

The actual calculation of the values of the functions for use in the tables is carried out by using infinite-series representations of the trigonometric functions. Although this topic is outside the scope of this book, the basic idea will be described briefly.

Using the methods of the calculus it may be shown that the sine and cosine functions may be represented by various infinite series of algebraic terms called power series. Examples of such representations are shown in the two formulas below:

$$\sin x = x - \frac{x^3}{3!} + \frac{x^5}{5!} - \frac{x^7}{7!} + \cdots \tag{4}$$

$$\cos x = 1 - \frac{x^2}{2!} + \frac{x^4}{4!} - \frac{x^6}{6!} + \cdots \tag{5}$$

These series continue indefinitely according to rules that should be rather obvious from a study of the first few terms. It should be emphasized that the series formulas shown above are valid only when x is measured in radians.

These series have a property known as **convergence.** This means that in each case the sum of a finite number of terms is approaching

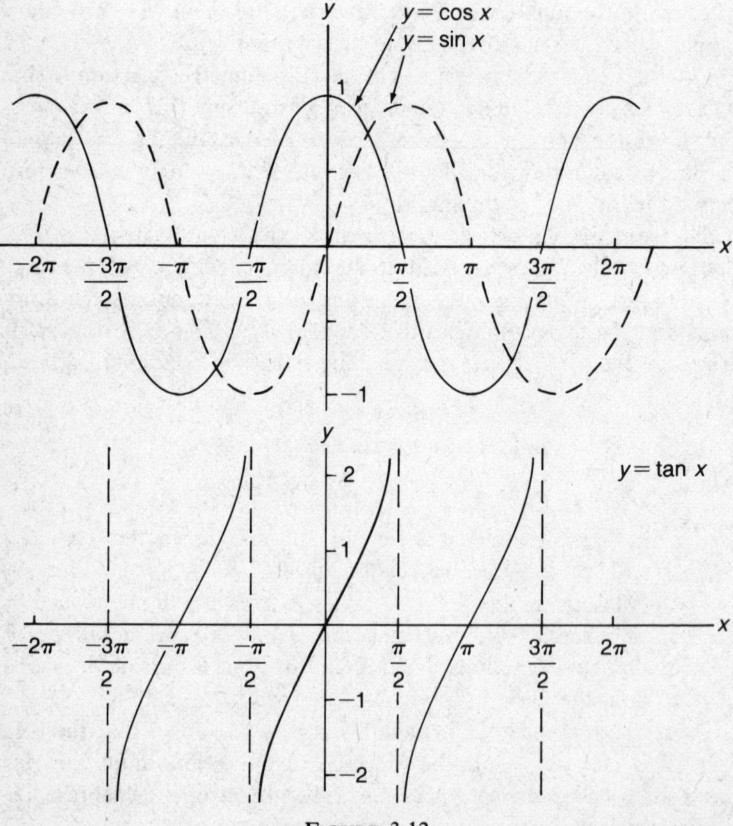

FIGURE 3.12

a limit in somewhat the same sense as a function was said to approach a limit in Section 2.3. The series (4) could be used to calculate the value of $\sin x$ to any desired degree of accuracy by substituting the specified value of x in the right-hand side of the equation and using a sufficient number of terms. The series (4) is particularly useful for calculating $\sin x$ for values of x close to 0.

By use of a table of the trigonometric functions, the graphs of these functions can be constructed in the usual way. Note that since the functions are periodic, the graphs repeat themselves over and over, so that a complete picture of the graph may be obtained by sketching it over an interval of one period.

Portions of the graphs of the sine, cosine, and tangent functions are shown in Figure 3.12.

3.12 Polar Coordinates

The trigonometric functions are useful in connection with a system of coordinates that may be employed as an alternative to the rectangular coordinate system, the **polar coordinate** system.

The polar system is based on a fixed point called the **pole** and a half-line* emanating from the pole called the **polar axis.** As in the rectangular system, a point in the plane is located by means of two numbers called coordinates, which are denoted by the letters r and θ. To locate a point in the plane whose polar coordinates are (r, θ), where $r > 0$, we proceed from the pole a distance r along the terminal side of an angle θ of which the polar axis forms the initial side, as shown in Figure 3.13.

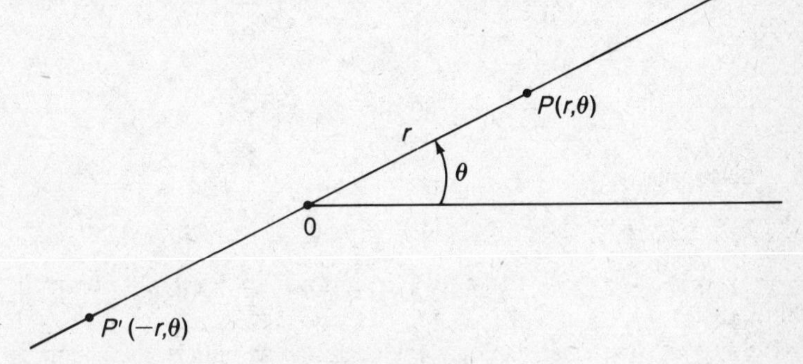

FIGURE 3.13

* A half-line is a line that originates at a point and extends indefinitely in one direction only.

A point for which $r < 0$ is located by proceeding a distance $|r|$ from the pole along the extension of the terminal side of θ. Such a point is also shown in Figure 3.13. Evidently the point $(0,\theta)$ is the pole regardless of the value of θ. The angle θ, which is measured in either degrees or radians, may assume the value 0 or any positive or negative value.

As a consequence of the above definition, it follows that any point may be designated by an unlimited number of different sets of polar coordinates. For example, the following sets of polar coordinates refer to the same point.

$$(4,30°),\quad (-4,210°),\quad (4,390°),\quad (-4,-150°)$$

If radian measure is employed, the same sets of polar coordinates are given by

$$\left(4,\frac{\pi}{6}\right),\quad \left(-4,\frac{7\pi}{6}\right),\quad \left(4,\frac{13\pi}{6}\right),\quad \left(-4,\frac{-5\pi}{6}\right)$$

In general, the coordinates (r, θ), $(-r, \theta + \pi)$, and $(r, \theta \pm 2n\pi)$, where n is an integer, refer to the same point.

It is frequently convenient to superimpose the polar system on the rectangular coordinate system. This is done by placing the pole at the origin of the rectangular system and making the polar axis coincide with the positive x axis. If this is done, a simple relationship between the rectangular and the polar coordinates of a point may be readily established. Referring to Figure 3.14, we readily see that

$$\cos\theta = \frac{x}{r} \qquad \text{so that } x = r\cos\theta$$

$$\sin\theta = \frac{y}{r} \qquad \text{so that } y = r\sin\theta$$

Furthermore,

$$x^2 + y^2 = r^2\cos^2\theta + r^2\sin^2\theta = r^2(\cos^2\theta + \sin^2\theta) = r^2$$

It may be shown that the above equations hold for all points in the plane.

By use of these relationships, the rectangular equations of curves in the plane may be easily transformed to polar form and vice-versa. For example, the equation of the unit circle, $x^2 + y^2 = 1$, becomes $r = 1$ in polar form and that for the straight line, $y = x$, becomes

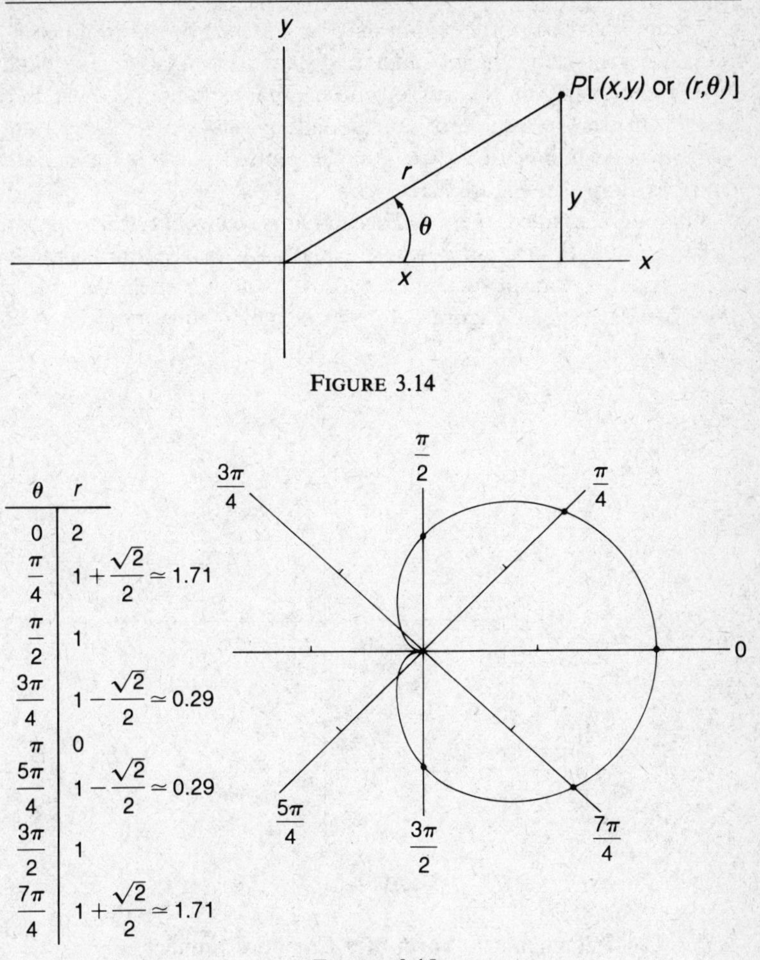

FIGURE 3.14

θ	r
0	2
$\frac{\pi}{4}$	$1+\frac{\sqrt{2}}{2} \simeq 1.71$
$\frac{\pi}{2}$	1
$\frac{3\pi}{4}$	$1-\frac{\sqrt{2}}{2} \simeq 0.29$
π	0
$\frac{5\pi}{4}$	$1-\frac{\sqrt{2}}{2} \simeq 0.29$
$\frac{3\pi}{2}$	1
$\frac{7\pi}{4}$	$1+\frac{\sqrt{2}}{2} \simeq 1.71$

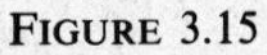
FIGURE 3.15

$\theta = \pi/4$. The polar equation $r = \sin\theta$ may be transformed to rectangular form by first multiplying both sides of the equation by r to obtain

$$r^2 = r\sin\theta$$

or

$$x^2 + y^2 = y$$

Using the technique of completing the square, this may be shown to be the equation of a circle with center at $(0,\frac{1}{2})$ and radius $\frac{1}{2}$.

The graph of a polar equation may be sketched by plotting points, just as is done using the rectangular system. For example, we shall sketch the graph of the curve whose polar equation is $r = 1 + \cos\theta$. We make a table of corresponding values of r and θ, plot the indicated points, and then join the plotted points by a smooth curve as shown in Figure 3.15.

This curve is called a **cardioid** because it is somewhat heart-shaped. It is an example of a curve whose equation is considerably simpler in polar form than in rectangular form. Another such example is the spiral shown in Figure 3.16, whose polar equation is $r = \theta$, $\theta \geq 0$.

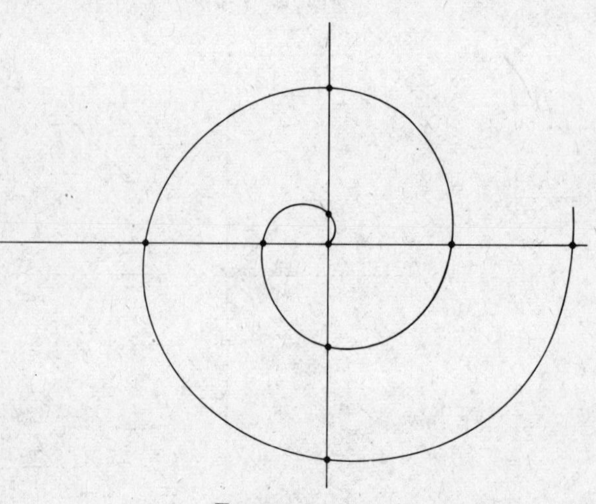

FIGURE 3.16

3.13 The Trigonometric Form of a Complex Number

In Section 1.10 it was shown that the complex number $a + bi$ could be alternatively designated as the ordered pair (a,b). This notation suggests a geometrical representation of a complex number as follows. With reference to a rectangular coordinate system, the complex number $a + bi$ corresponds to the point (a,b), as indicated in Figure 3.17.

Under this representation, it is clear that the real numbers will correspond to points on the x axis and the pure imaginary numbers, to points on the y axis, so that it is appropriate to call the x axis the real axis and the y axis the imaginary axis. The distance from the

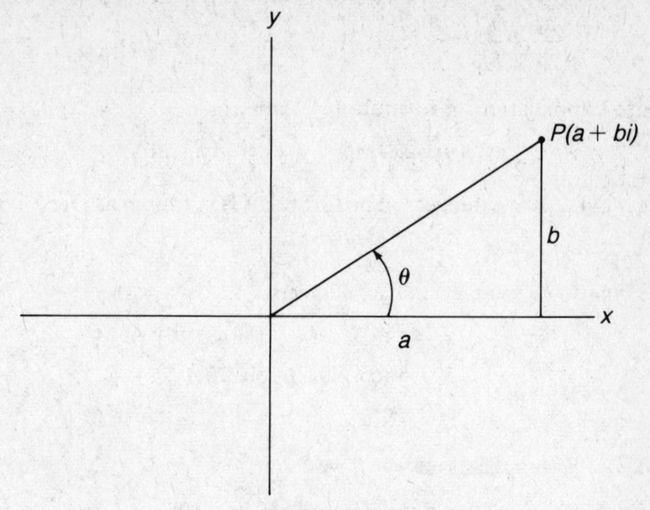

FIGURE 3.17

origin to the point corresponding to the number $a + bi$, indicated by the letter r, is called the **modulus,** or **absolute value,** of the complex number. From Figure 3.17 it is readily seen that

$$r = \sqrt{a^2 + b^2}$$

The angle θ, as indicated in Figure 3.17, is called the **amplitude,** or the **argument,** of the complex number. Furthermore, we see that $a = r \cos \theta$ and $b = r \sin \theta$, so that we have

$$a + bi = r \cos \theta + (r \sin \theta)i$$

or

$$a + bi = r(\cos \theta + i \sin \theta)$$

The right-hand side of the above equation, called the trigonometric form of the complex number $a + bi$, is very useful in performing certain algebraic operations. If

$$a_1 + b_1 i = r_1 (\cos \theta_1 + i \sin \theta_1)$$

and

$$a_2 + b_2 i = r_2 (\cos \theta_2 + i \sin \theta_2)$$

then it can be shown, by use of the addition formulas for the trigonometric functions, that

$$(a_1 + b_1 i)(a_2 + b_2 i) = r_1 r_2 [\cos (\theta_1 + \theta_2) + i \sin (\theta_1 + \theta_2)] \quad (1)$$

and $$\frac{a_1 + b_1 i}{a_2 + b_2 i} = \frac{r_1}{r_2} [\cos (\theta_1 - \theta_2) + i \sin (\theta_1 - \theta_2)] \tag{2}$$

Repeated application of formula (1) leads to

$$(a_1 + b_1 i)^n = r_1{}^n(\cos n\theta_1 + i \sin n\theta_1) \tag{3}$$

where n is a positive integer. The formula (3) is known as Demoivre's theorem.

Example 1 Given the complex numbers

$$z_1 = 3 (\cos 40° + i \sin 40°)$$
$$z_2 = 5 (\cos 20° + i \sin 20°)$$

find $z_1 z_2$.

Solution Using the formula (1), we have

$$\begin{aligned} z_1 z_2 &= 15 (\cos 60° + i \sin 60°) \\ &= \frac{15}{2} + \frac{15\sqrt{3}}{2} i \\ &= \frac{15}{2} (1 + i\sqrt{3}) \end{aligned}$$

Example 2 Given the complex numbers

$$z_1 = 12 (\cos 80° + i \sin 80°)$$
$$z_2 = 6 (\cos 50° + i \sin 50°)$$

find z_1/z_2.

Solution Use formula (2)

$$\begin{aligned} \frac{z_1}{z_2} &= 2 (\cos 30° + i \sin 30°) \\ &= \sqrt{3} + i \end{aligned}$$

Example 3 Find $(1 + i)^{10}$

Solution First we transform the complex number $1 + i$ to trigonometric form. Since $a = 1$ and $b = 1$, we have

$$r = \sqrt{a^2 + b^2} = \sqrt{1^2 + 1^2} = \sqrt{2}$$

Furthermore, the equations $a = r \cos \theta$ and $b = r \sin \theta$ become

$$1 = \sqrt{2} \cos \theta \text{ and } 1 = \sqrt{2} \sin \theta$$

so that $\cos\theta = \sin\theta = 1/\sqrt{2}$. Hence we may take $\theta = 45°$ (or $45° + 2n\pi$, where n is an integer). Therefore, $1 + i = \sqrt{2}\,(\cos 45° + i \sin 45°)$. Hence we have, using formula (3),

$$\begin{aligned}(1+i)^{10} &= (\sqrt{2})^{10}\,(\cos 450° + i \sin 450°)\\ &= (\sqrt{2})^{10}\,(\cos 90° + i \sin 90°)\\ &= 32i\end{aligned}$$

The fact that Demoivre's theorem can be shown to hold for fractional exponents can be used to find the roots of a complex number. First of all it may be noted that from the fact that the sine and cosine functions are periodic with period 2π it follows that

$$\begin{aligned}&r(\cos\theta + i \sin\theta)\\ &\qquad = r\cos(\theta + k \cdot 360°) = i \sin(\theta + k \cdot 360°]\end{aligned}$$

where k is an integer, and hence

$$\begin{aligned}&[r(\cos\theta + i \sin\theta)]^{1/n}\\ &\qquad = r^{1/n}\left[\cos\left(\frac{\theta + k \cdot 360°}{n}\right) + i \sin\left(\frac{\theta + k \cdot 360°}{n}\right)\right]\end{aligned}$$

As noted, k may be any integer, and we may obtain the n distinct nth roots of a complex number that has been expressed in trigonometric form by allowing k to assume the values $0, 1, 2, \ldots, n - 1$. Allowing k to assume additional values would result in simply duplicating roots already obtained. For example, if $k = n$

$$\frac{\theta + n \cdot 360°}{n} = \frac{\theta}{n} + 360°$$

so that the same root that was calculated by letting $k = 0$ would be obtained again.

For example, let us obtain the cube roots of the real number -27. In the form $a + bi$ this number appears as $-27 + 0i$, and its trigonometric form is readily seen to be

$$27(\cos 180° + i \sin 180°)$$

Hence

$$\begin{aligned}(-27 + 0i)^{1/3} &= [27(\cos 180° + i \sin 180°]^{1/3}\\ &= (27)^{1/3}\left[\left(\cos\frac{180° + k \cdot 360°}{3}\right.\right.\\ &\qquad \left.\left. + i \sin\frac{180° + k \cdot 360°}{3}\right)\right]\end{aligned}$$

$$= 3[\cos(60° + k \cdot 120°) + i \sin (60° + k \cdot 120°)]$$

$$k = 0, 1, 2$$

We then obtain the following roots, using the specified values of k:

$$k = 0 \qquad 3(\cos 60° + i \sin 60°) = \frac{3}{2} + 3\frac{\sqrt{3}}{2}i$$

$$k = 1 \qquad 3(\cos 180° + i \sin 180°) = -3$$

$$k = 2 \qquad 3(\cos 300° + i \sin 300°) = \frac{3}{2} - \frac{3\sqrt{3}}{2}i$$

3.14 The Inverse Trigonometric Functions

The term, "inverse trigonometric function," is actually a misnomer since the trigonometric functions actually have no inverse. In Section 2.4 it was pointed out that to have an inverse, a function must be one-to-one, i.e., the same value in the range must not correspond to more than one value in the domain. Hence the function $y = \sin x$, for example, is not a one-to-one function since the value $y = 1/2$ corresponds to $x = \pi/6$, $5\pi/6$, $13\pi/6$, and an infinite number of additional positive and negative values of x. The same situation exists in the case of the other trigonometric functions.

Although, as indicated above, the trigonometric functions have no inverses, we can readily define a group of functions that do possess inverses by limiting the domain of each of the trigonometric functions. For the sine, cosine, and tangent functions, we define the following functions, the names of which are capitalized to distinguish them from the ordinary trigonometric functions:

$$\text{Sin } x = \sin x \qquad -\frac{\pi}{2} \leq x \leq \frac{\pi}{2} \tag{1}$$

$$\text{Cos } x = \cos x \qquad 0 \leq x \leq \pi \tag{2}$$

$$\text{Tan } x = \tan x \qquad -\frac{\pi}{2} < x < \frac{\pi}{2} \tag{3}$$

These functions may readily be seen to be one-to-one. The inverses of these three functions are designated by arcsin x, arccos x, and arctan x, respectively.* The graphs, arcsin x, arccos x, and arctan x, are shown in Figure 3.18.

* These functions may also be designated by $\sin^{-1} x$, $\cos^{-1} x$, and $\tan^{-1} x$, respectively.

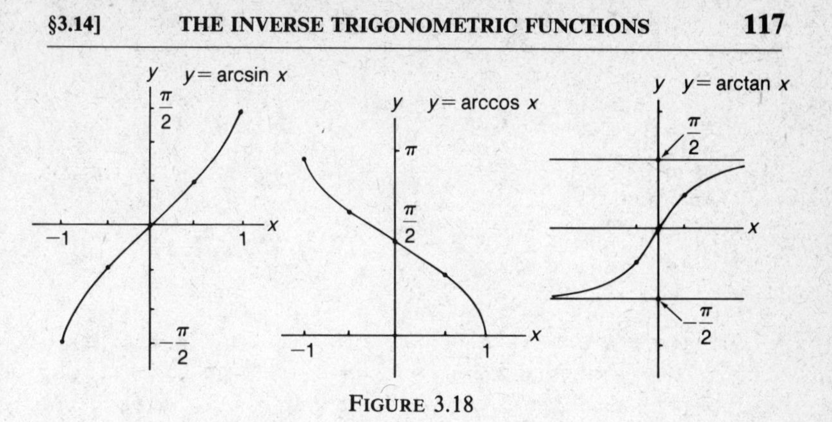

FIGURE 3.18

A study of the graph of the cosine function (Figure 3.12) will show why the interval $[0,\pi]$ was chosen for the domain of the function Cos x rather than the interval $[-\pi/2,\pi/2]$, which was used for the function Sin x. Clearly the function Sin x is one-to-one in the interval $[-\pi/2,\pi/2]$, but Cos x is not.

Finally, it should be noted that since in forming the inverse of a given function, we interchange the range and the domain of the function, we have the following:

Function	Domain	Range
arcsin x	$[-1,1]$	$\left[-\frac{\pi}{2},\frac{\pi}{2}\right]$
arccos x	$[-1,1]$	$[0,\pi]$
arctan x	$(-\infty,\infty)$	$\left(-\frac{\pi}{2},\frac{\pi}{2}\right)$

In view of the definitions of the inverse functions, the following statements can be made:

$$\arcsin x = \text{the number in } \left[-\frac{\pi}{2},\frac{\pi}{2}\right] \text{ whose sin is } x$$

$$\arccos x = \text{the number in } [0,\pi] \text{ whose cos is } x$$

$$\arctan x = \text{the number in } \left(-\frac{\pi}{2},\frac{\pi}{2}\right) \text{ whose tan is } x$$

Keeping the above information in mind, we obtain the following values of the inverse trigonometric functions:

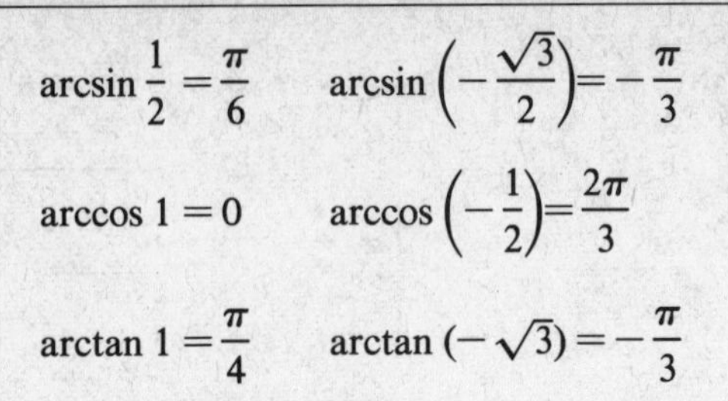

$$\arcsin\frac{1}{2}=\frac{\pi}{6} \qquad \arcsin\left(-\frac{\sqrt{3}}{2}\right)=-\frac{\pi}{3}$$

$$\arccos 1=0 \qquad \arccos\left(-\frac{1}{2}\right)=\frac{2\pi}{3}$$

$$\arctan 1=\frac{\pi}{4} \qquad \arctan(-\sqrt{3})=-\frac{\pi}{3}$$

The inverse trigonometric functions arise naturally in many problems in the calculus and are useful in certain applications. For example, the shaded area shown in the graph of $y = 1/(1 + x^2)$ shown in Figure 3.19 may be shown to be arctan b − arctan a.

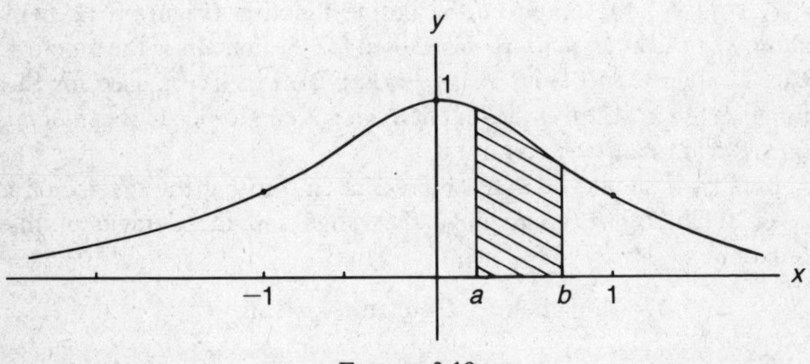

FIGURE 3.19

In some cases it is useful to be able to change certain expressions involving both the trigonometric and inverse trigonometric functions to algebraic form. Consider, for example, the expression cos (arcsin x). We are seeking the cosine of an angle whose sine is x. But $\cos\theta = \pm\sqrt{1-\sin^2\theta}$. Hence if we let $\theta = \arcsin x$, since sin (arcsin x) = x we have

$$\cos(\arcsin x)=\cos\theta=\sqrt{1-x^2}$$

where the positive sign has been chosen for the radical since the range of arcsin x is $[-\pi/2,\pi/2]$, and the cosine is non-negative in this interval.

For another example, consider the process of simplifying the equation

$$\arctan x+\arctan y=\arctan xy$$

Recalling the formula

$$\tan(\alpha+\beta)=\frac{\tan\alpha+\tan\beta}{1-\tan\alpha\tan\beta}$$

and letting $\arctan x=\alpha$ and $\arctan y=\beta$, we take the tangent of both sides of the given equation. Since $\tan(\arctan x)=x$, etc., it is readily seen that the given equation reduces to

$$\frac{x+y}{1-xy}=xy$$

3.15 The Hyperbolic Functions

Certain combinations of exponential functions have important and interesting properties that justify their being given a separate designation. They are called the hyperbolic functions. Their basic definitions are those of the **hyperbolic sine** and the **hyperbolic cosine functions,** which are designated as $\sinh x$ and $\cosh x$, respectively. They are defined as follows:

$$\sinh x=\frac{e^x-e^{-x}}{2} \tag{1}$$

$$\cosh x=\frac{e^x+e^{-x}}{2} \tag{2}$$

The similarity between the names of these functions and those of the trigonometric functions is due to the remarkable analogies between the properties of these functions and those of the trigonometric functions. For example, corresponding to the trigonometric identity $\sin^2 x+\cos^2 x=1$ we have, for the hyperbolic functions,

$$\cosh^2 x-\sinh^2 x=1 \tag{3}$$

since

$$\cosh^2 x-\sinh^2 x=\left(\frac{e^x+e^{-x}}{2}\right)^2-\left(\frac{e^x-e^{-x}}{2}\right)^2$$

$$=\frac{e^{2x}+2+e^{-2x}}{4}-\frac{e^{2x}-2+e^{-2x}}{4}$$

$$=\frac{e^{2x}+2+e^{-2x}-e^{2x}+2-e^{-2x}}{4}=1$$

Equation (3) accounts in part for the use of the term "hyperbolic" in connection with these functions. In this there is an analogy to the designation of the trigonometric functions as circular functions.

If a point (x,y) on a curve is determined by the equations

$$x = a \cos t \qquad y = a \sin t$$

where a is a constant, then

$$\begin{aligned} x^2 + y^2 &= a^2 \cos^2 t + a^2 \sin^2 t \\ &= a^2 (\cos^2 t + \sin^2 t) = a^2 \end{aligned}$$

so that the curve is seen to be a circle with the center at the origin and radius $|a|$. Furthermore, equations of this type, in which x and y are given in terms of a third variable, are said to be **parametric** equations and the third variable is called a **parameter.**

In the same way, if the points on a curve are determined by the parametric equations

$$x = a \cosh t \qquad y = a \sinh t$$

then

$$\begin{aligned} x^2 - y^2 &= a^2 \cosh^2 t - a^2 \sinh^2 t \\ &= a^2 (\cosh^2 t - \sinh^2 t) = a^2 \end{aligned}$$

The curve is seen to be an equilateral hyperbola with center at the origin and transverse and conjugate axes each of length $2|a|$.

There are four other hyperbolic functions, which are defined as follows, in a manner which makes use of corresponding relationships among the trigonometric functions.

$$\tanh x = \frac{\sinh x}{\cosh x} = \frac{e^x - e^{-x}}{e^x + e^{-x}} \tag{4}$$

$$\operatorname{csch} x = \frac{1}{\sinh x} = \frac{2}{e^x - e^{-x}} \tag{5}$$

$$\operatorname{sech} x = \frac{1}{\cosh x} = \frac{2}{e^x + e^{-x}} \tag{6}$$

$$\coth x = \frac{\cosh x}{\sinh x} = \frac{e^x + e^{-x}}{e^x - e^{-x}} \tag{7}$$

Using these definitions, we can establish other identities similar to those that hold for the trigonometric functions, for example,

$$1 - \tanh^2 x = \operatorname{sech}^2 x \tag{8}$$

$$\coth^2 x - 1 = \operatorname{csch}^2 x \tag{9}$$

$$\sinh(x + y) = \sinh x \cosh y + \cosh x \sinh y \tag{10}$$

The hyperbolic functions have even and odd properties similar to those which hold for the trigonometric functions. For example,

$$\sinh(-x) = \frac{e^{-x} - e^{x}}{2} = -\sinh x$$

so that the hyperbolic sine is an odd function. On the other hand,

$$\cosh(-x) = \frac{e^{-x} + e^{x}}{2} = \cosh x$$

so that the hyperbolic cosine is an even function. The function $\tanh x$ is readily seen to be an odd function. Through the reciprocal relationships it is evident that $\operatorname{csch} x$ and $\coth x$ are odd, whereas $\operatorname{sech} x$ is even.

An interesting application of the hyberbolic cosine function involves the curve known as the **catenary,** whose equation is of the form $y = a \cosh x/a$, where a is a constant. It can be shown that a flexible chain suspended at two points hangs in a curve of this type. [See Figure 3.20 (b).] The graphs of $\sinh x$, $\cosh x$, and $\tanh x$ are shown in Figure 3.20.

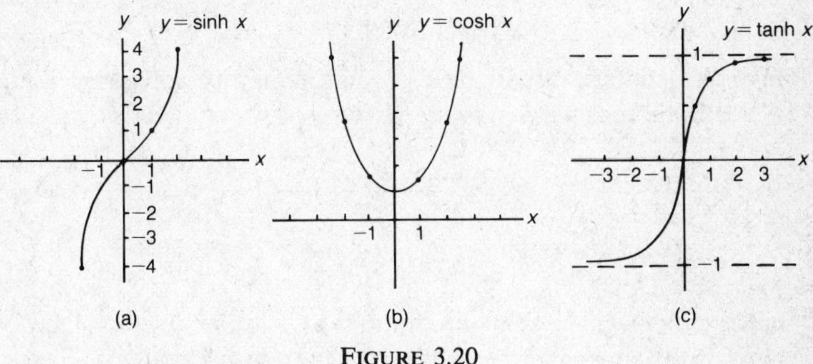

FIGURE 3.20

3.16 The Inverse Hyperbolic Functions

All of the hyperbolic functions with the exceptions of $\cosh x$ and $\operatorname{sech} x$ are one-to-one functions, hence they have inverses, denoted by $\sinh^{-1} x$, $\tanh^{-1} x$, etc. It should be noted that in this notation the -1 is not an exponent.

The functions cosh x and sech x, since they are even, are obviously not one-to-one and hence do not have inverses. However, as was the case with the trigonometric functions, we may obtain inverses for modifications of these functions obtained by limiting their domains. These functions are

$$\text{Cosh } x = \cosh x \qquad x \geq 0$$

$$\text{Sech } x = \text{sech } x \qquad x \geq 0$$

These functions have inverses, designated as $\cosh^{-1} x$ and $\text{sech}^{-1} x$.

The inverse hyperbolic functions can be expressed in logarithmic form. This seems reasonable since the hyperbolic functions are exponential in form, and, as we have seen, the inverse of the exponential function is the logarithmic function.

For an illustration of this property consider the function $y = \sinh^{-1} x$. This is equivalent to the equation

$$x = \sinh y = \frac{e^y - e^{-y}}{2}$$

It may be written as

$$e^y - e^{-y} - 2x = 0$$

Multiplying by e^y, we obtain

$$e^{2y} - 2xe^y - 1 = 0$$

This is a quadratic equation in e^y, and it may be solved by using the quadratic formula where $a = 1$, $b = -2x$, $c = -1$:

$$e^y = \frac{2x \pm \sqrt{4x^2 + 4}}{2}$$

$$= x \pm \sqrt{x^2 + 1}$$

The negative sign in front of the radical may be deleted since e^y cannot be negative and $x - \sqrt{x^2 + 1}$ is negative for all x. Hence, taking the natural logarithm of both sides, we obtain

$$y = \sinh^{-1} x = \ln(x + \sqrt{x^2 + 1})$$

A similar procedure may be used to express the other inverse hyperbolic functions in logarithmic form.

Problems—Chapter 3

1. Find the value of each of the following:

(a) $\cos 150°$
(b) $\tan 120°$
(c) $\sin 135°$
(d) $\sin 210°$
(e) $\cos 240°$
(f) $\tan 225°$
(g) $\tan 330°$
(h) $\sin 300°$
(i) $\cos 315°$
(j) $\sin 510°$
(k) $\cos 1050°$
(l) $\tan 585°$
(m) $\sin(-240°)$
(n) $\cos(-330°)$
(o) $\tan(-480°)$

2. Use trigonometric identities to simplify each of the following expressions:

(a) $\dfrac{1}{1+\cos\alpha}+\dfrac{1}{1-\cos\alpha}$
(b) $\sin^4\alpha+\sin^2\alpha\cos^2\alpha+\cos^2\alpha$
(c) $\dfrac{1+\cot^2\alpha}{\sec^2\alpha}$
(d) $\cot^2\alpha\sec^2\alpha-\cot^2\alpha$

3. Find the primary-angle solutions of each of the following equations:

(a) $2\cos\alpha+1=0$
(b) $2\sin^2\alpha-3\sin\alpha+1=0$
(c) $\cos\alpha\tan\alpha+\cos\alpha=0$
(d) $4\cos^2\alpha-3=0$

4. Sketch the graph of each of the following polar equations:

(a) $r=1-\sin\theta$
(b) $r=3\cos\theta$

5. Given the complex numbers $z_1=4(\cos 20°+i\sin 20°)$ and $z_2=5(\cos 25°+i\sin 25°)$, express the product z_1z_2 in the form $a+bi$.

6. Given the complex numbers $z_1=8(\cos 165°+i\sin 165°)$ and $z_2=2(\cos 15°+i\sin 15°)$, express the quotient z_1/z_2 in the form $a+bi$.

7. Find the three cube roots of 64.

8. Evaluate, using radian measure:

(a) $\arcsin(-\frac{1}{2})$
(b) $\arccos-(\sqrt{3}/2)$
(c) $\arctan 0$
(d) $\arcsin 1$
(e) $\arccos(-1)$
(f) $\arctan(-1/\sqrt{3})$
(g) $\arcsin(\sin\pi/3)$
(h) $\arccos(\cos 7\pi/6)$
(i) $\arctan(\tan 3\pi/4)$

9. Using the definitions of $\sinh x$ and $\cosh x$, prove the identity

$$\sinh(x+y)=\sinh x\cosh y+\cosh x\sinh y$$

4

ELEMENTARY PROBABILITY

4.1 Introduction

We live in a world of uncertainty. The outcome of most future events is unknown to us. Probability is a branch of mathematics that deals with the assessment of the likelihood of future events by assigning numerical measures to their likelihood.

The origins of the theory of probability may be traced to the seventeenth century, when certain questions arose concerning the gambling games popular in the fashionable society of France. Several prominent mathematicians of the day, including Blaise Pascal and Pierre Fermat, addressing themselves to these problems, furnished solutions that constituted the beginnings of the theory of probability.

Today, however, the theory of probability has gone far beyond its somewhat trivial origins. It now forms one of the most important branches of both theoretical and applied mathematics. Its impact is felt in business and industry, in sociology, in medicine, and in innumerable other aspects of modern life.

4.2 Sample Spaces and Events

We begin our study of the elementary theory of probability with the definition of an **experiment.** An experiment is any action that results in an observation or a set of observations. For example, the flipping of a coin is an experiment, the observed result of which is either a head or a tail. Another example of an experiment is the selection of a group of people from a larger group and the recording of their weights to the nearest pound, the result being a set of integers.

The second basic idea in the development of elementary probability theory is the concept of a sample space. A **sample space** for a given experiment is simply the set of all possible outcomes of the experiment according to some classification of the outcomes. For example, consider the experiment of tossing a coin twice. If we are concerned only with whether the coin falls head or tail on each toss and if the result of each toss is indicated by H for head and T for tail so that HT indicates the outcome, head on the first toss and tail on the second, then a sample space S for this experiment can be represented as follows:

$$S = \{\text{HH,HT,TH,TT}\}$$

However, other sample spaces can be devised in connection with this experiment. For example, if the outcomes are classified according to the number of heads that occur, a sample space for the experiment is $S = \{0,1,2\}$. Furthermore, if the outcomes are classified by whether the results of the two tosses are the same (s) or different (d), then a sample space for the experiment is $S = \{\text{s,d}\}$. In view of these considerations, we speak of *a* sample space rather than *the* sample space for a given experiment.

Sample spaces containing an infinite number of elements arise in connection with certain experiments. For example, consider the experiment of conducting a series of tosses of a coin until a head occurs, and define the outcome of the experiment as the number of the toss on which the first head occurs. Evidently the number of possible outcomes is unlimited and the sample space could be designated as $S = \{1,2, \ldots\}$. The number of outcomes in this case is said to be countably infinite since there is a one-to-one correspondence between the set of outcomes and the set of positive integers.

Sample spaces containing a finite or countably infinite number of elements are said to be **discrete,** whereas those which contain an uncountably infinite number are called **continuous.** In this chapter we shall confine our discussion to discrete sample spaces and mainly to those containing a finite number of elements. All sample spaces will be assumed to be finite unless otherwise stated.

Our next definition in developing the theory of probability is for an event. An **event** is defined as any subset of a sample space. For example, an event A associated with the sample space $S = \{\text{HH,HT,TH,TT}\}$ referred to above is the set $A = \{\text{HH,HT,TH}\}$. This event might be defined verbally as follows: at least one head occurs. Furthermore, an event consisting of a single element such as $\{\text{H,T}\}$ is called an **elementary event.** An event A is said to occur if one of the outcomes contained in A occurs.

It is clear that many other events could be formed from this same sample space. This leads to the following question. Given a sample space containing N elements, how many different events can be associated with it? The answer can readily be seen to be 2^N. For example, there are 2^4 or 16 events associated with the sample space referred to above in connection with the coin tossing experiment.

The reasoning leading to the formula given above is relatively simple. In forming a given subset of a sample space one considers each element and either includes it in the event or does not include it. That is, there are two ways of disposing of each element of the sample space. There are therefore 2^N ways of disposing of all of the events of the sample space. It should be noted that in our counting process we have included an event in which we have used none of the elements of the sample space, or, in set language, we have formed an empty set. This event is called the **impossible event** since it will never occur. Its inclusion may seem unrealistic, but it is a mathematical convenience in the development of the theory. Evidently, if the number of elements in the sample space is large, the number of events associated with it is extremely large. For example, the number of events associated with a sample space containing 20 elements is 2^{20}, or 1,048,576.

We often have occasion to consider combinations of two or more events. In this connection the set notation previously employed is extremely useful, and we shall employ it as follows. By $A \cup B$ we shall mean the occurrence of either the event A or the event B or both. By $A \cap B$ we shall mean the occurrence of both of the events A and B. The use of these symbols is, of course, in perfect accord with the meanings of the union and intersection symbols in the algebra of sets. Furthermore, these symbols are sometimes used in conjunction with more than two events. For example, $A \cap B \cap C$ refers to the occurrence of each of the events A, B, and C.

Two or more events are said to be **mutually exclusive** if they contain no elementary events in common. This corresponds to the idea of disjoint sets. For example, in connection with the coin tossing experiment the events (HH,HT) and (TH,TT) are mutually exclusive, i.e., it is impossible for both of them to occur.

4.3 Probability Functions

We have now laid the necessary groundwork for the definition of a probability function, which will enable us to assign a number called a probability to each event in a given sample space. A

probability function P is a set function—i.e., the elements of its domain are sets rather than numbers—that satisfies the following axioms, where A is any event in a sample space S:

(1) $P(A) \geq 0$

(2) $P(S) = 1$

(3) If A and B are mutually exclusive events, then $P(A \cup B) = P(A) + P(B)$.

The number $P(A)$ is called the probability of the event A.

It should be emphasized that the three statements above are axioms and therefore require no proof. However, we shall use them to prove various simple theorems to be used in turn in the calculation of probabilities.

It should be pointed out that another set of axioms might have been used to define a probability function. However, the question that arises in connection with the choice of a set of axioms is whether they constitute the definition of a function that will be useful in the real world. As we shall see, the axioms we have selected meet this standard.

First we prove a few basic theorems.

Theorem I. If $A_1, A_2, \ldots, A_n$ are n mutually exclusive events in a sample space S, then

$$P(A_1 \cup A_2 \cup \cdots \cup A_n) = P(A_1) + P(A_2) + \cdots + P(A_n)$$

Proof By axiom (3)

$$P(A_1 \cup A_2) = P(A_1) + P(A_2)$$

Denote the event $A_1 \cup A_2$ by the symbol A_{12}. Then the events A_{12} and A_3 are mutually exclusive. Hence, by axiom (3)

$$P(A_{12} \cup A_3) = P(A_{12}) + P(A_3)$$

Hence $P(A_1 \cup A_2 \cup A_3) = P(A_{12} \cup A_3) = P(A_1) + P(A_2) + P(A_3)$

Clearly the same line of reasoning can be extended to any finite number of mutually exclusive events.

Theorem II. The probability of an event that contains no elements of S (the impossible event) is 0. The probability of any event A that contains at least one element of S is the sum of the probabilities of the elementary events in A.

Proof If we designate the impossible event by I, then S and I are mutually exclusive events. Hence, by axiom (3)

$$P(S \cup I) = P(S) + P(I)$$

But $$S \cup I = S$$

Hence $$P(S) + P(I) = P(S)$$

Therefore $$P(I) = 0$$

If A contains at least one element of S, then denote the elementary events in A by $E_1, E_2, \ldots, E_n$.

Hence $$A = E_1 \cup E_2 \cup \cdots \cup E_n$$

by Theorem I

$$P(A) = P(E_1 \cup E_2 \cup \cdots \cup E_n) = P(E_1) + P(E_2) + \cdots + P(E_n)$$

Theorem III. For any event A

$$0 \leq P(A) \leq 1$$

Proof By axiom (1), $P(A) \geq 0$. By axiom (2), $P(S) = 1$. Since A is a subset of S, we have, by Theorem II, $P(A) \leq P(S) = 1$.

We are now in a position to assess the suitability of the axioms we have used to define a probability function. We note that to most people the probability of an event denotes the fraction of the time the event will occur in the long run. This fraction will, of course, be a number between 0 and 1, where the probability 0 denotes an impossible event and the probability 1 denotes a certain event. This is in accord with Theorem III, which follows from axioms (1) and (2).

Next we note that when an experiment is performed, one of the outcomes is certain to occur. This agrees with axiom (2). Finally we note that if an event occurs a given fraction of the time and another mutually exclusive event occurs another fraction of the time, then one or the other event occurs a fraction of the time equal to the sum of the two fractions. For example, if it rains one-quarter of the time and if it snows one-eighth of the time, then it either rains or snows three-eighths of the time. This is in accord with axiom (3).

4.4 The Calculation of Probabilities

We are now in a position to calculate the probability of any event A, provided that we know the probability of each of the elementary events contained in A. However, none of our axioms or theorems tells us exactly how to assign these probabilities. We do know that each of the probabilities assigned must be a number between 0 and 1 and that the sum of the probabilities assigned to all of the elements of the sample space must be equal to 1.

A frequently employed device is to assume that each of the elementary outcomes of the experiment is equally likely to occur, which is often a reasonable assumption. For example, if we have no evidence to the contrary, it seems reasonable to assume that if a coin is tossed, a head and a tail are equally likely to occur. Hence we would assign the probability $\frac{1}{2}$ to each of these outcomes. Similarly, if a single die that is assumed to be balanced is thrown and if the outcome is defined to be the number of spots on the side of the die that ends face up, then each outcome is assigned the probability $\frac{1}{6}$. In general, if a sample space has N elements and if it seems reasonable to assign equal probabilities to each element, then we would, in accordance with axiom (2), assign to each the probability $1/N$.

This discussion leads naturally to a simple but important theorem that enables us to calculate the probabilities of events of a certain type.

Theorem IV. If an experiment has N equally likely outcomes, and if n of these comprise the subset of S that constitutes the event A, then $P(A) = n/N$.

Proof By Theorem II, the probability of A is the sum of the probabilities of the elementary events in A. But since these are equally likely, each has probability $1/N$. Therefore,

$$P(A) = \overbrace{\frac{1}{N} + \frac{1}{N} + \cdots + \frac{1}{N}}^{n \text{ terms}} = \frac{n}{N}$$

We shall now illustrate Theorem IV with two simple examples.

Example 1 A single card is drawn from a thoroughly shuffled full deck. What is the probability that the card is a heart?

Solution Since there are 52 cards in the deck and 13 of them are hearts, the probability of drawing a heart is 13/52, or 1/4. Note that the requirement of *equal likelihood* of the outcomes is ensured by the shuffling of the cards.

Example 2 Two dice are thrown. What is the probability that their sum will be a 7?

Solution Since each die may fall in 6 ways, there are $6 \times 6 = 36$ ways in which both dice can fall, and if the dice are balanced, each of these outcomes is equally likely. In order to complete the solution it is necessary to count the number of ways in which a sum of 7 can be obtained. If the dice are identified as die 1 and die 2, we can readily see that the outcomes with sums of 7 can occur in the following ways:

Die 1	*Die 2*
6	1
5	2
4	3
3	4
2	5
1	6

There are therefore 6 ways in which a sum of 7 can occur. Therefore, the probability of throwing 7 is $\frac{7}{36}$, or $\frac{1}{6}$.

Before stating the next theorem, we introduce the concept of complementary events. If A is an event in a sample space S, then the event that consists of all of the elements of S not in A is denoted by the symbol A', and the events A and A' are called **complementary events.** Clearly A and A' are mutually exclusive events, and when the experiment is performed, either A or A' must occur. The event A' might be called the nonoccurrence of A.

We can now prove the following simple but useful theorem.

Theorem V. $P(A) = 1 - P(A')$

Proof Clearly $S = A \cup A'$. Hence since A and A' are mutually exclusive events, we have, using axiom (3),

$$P(S) = P(A \cup A') = P(A) + P(A')$$

Therefore, by axiom (2),

$$P(A) + P(A') = 1$$

and the theorem follows.

The above result is intuitively obvious. For example, if it rains one-quarter of the time, then evidently it does not rain three-quarters of the time.

4.5 Basic Laws of Probability; Conditional Probability

Many probability problems involve combinations of two or more events, and in this connection there are two basic laws that we shall now develop and use: the addition law and the multiplication law. We shall consider **the addition law** first.

Theorem VI. $P(A \cup B) = P(A) + P(B) - P(A \cap B)$

This theorem may be stated verbally as follows. The probability that either event A or event B, or possibly both A and B, will occur equals the probability of A plus the probability of B minus the probability that both A and B will occur.

Proof Consider the Venn diagrams shown in Figure 4.1. The dots represent the elementary events in A and B.

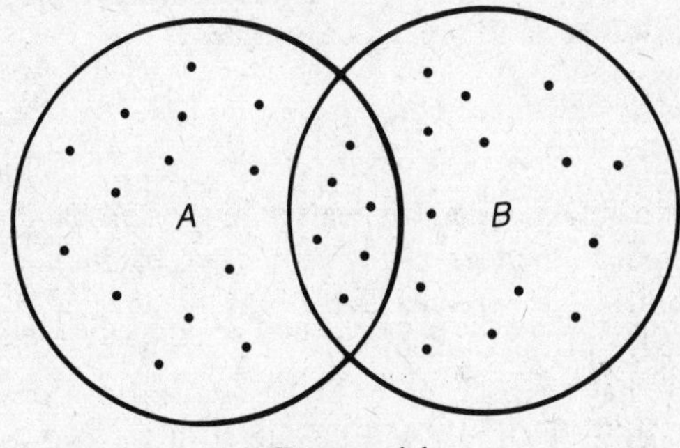

FIGURE 4.1

By Theorem II, $P(A)$ is obtained by adding the probability associated with each of the dots in A; similarly for $P(B)$. But in adding $P(A)$ and $P(B)$, we have added the probabilities associated with the dots common to A and B twice. These dots constitute the set $A \cap B$. Therefore $P(A \cap B)$ must be subtracted from the sum $P(A) + P(B)$. Hence the theorem has been proved.

The addition law can readily be extended to cases involving more than two events. For example, for three events A, B, and C it can be shown that

$$P(A \cup B \cup C) = P(A) + P(B) + P(C) - P(A \cap B) - P(A \cap C) - P(B \cap C) + P(A \cap B \cap C)$$

Example 1 A single card is drawn from a full deck. What is the probability that the card will be either a heart or a king?

Solution The probability that a single card drawn from a full deck will be a heart is 13/52. The probability that it will be a king is 4/52. The probability that the card will be both a king and heart is 1/52. Therefore,

$$P = \frac{13}{52} + \frac{4}{52} - \frac{1}{52} = \frac{16}{52} = \frac{4}{13}$$

In the special case in which A and B are mutually exclusive events, $P(A \cap B) = 0$ since $A \cap B$ is the impossible event. Hence Theorem VI reduces to $P(A \cup B) = P(A) + P(B)$. However, this is seen to be simply axiom (3) in the definition of a probability function. It will be recalled that in Theorem I this formula has been extended to the case of n mutually exclusive events.

Example 2 If a single card is drawn from a full deck, what is the probability that it will be either a heart or a spade?

$$P = \frac{1}{4} + \frac{1}{4} = \frac{1}{2}$$

In order to obtain the **multiplication law** for probability, it is necessary to introduce the idea of conditional probability. In some cases the probability of a specified event may be affected by the state of our knowledge of the outcome of the experiment. For example, it seems clear that the probability that a single card drawn at random from a full deck is a heart will be altered if we know that the card drawn is red. A probability of an event that is calculated in the light of the known occurrence of some other event is called a **conditional probability.** The symbol $P(B|A)$ is used to denote the probability of the event B, given that the event A has occurred.

We shall now obtain a formula for computing $P(B|A)$. In the discussion that follows we shall assume that $P(A) \neq 0$. A diagram of the sample space S and the events A and B is shown in Figure 4.2.

We desire a formula for the occurrence of B, when we know that A has occurred. Since A is known to have occurred, we redefine the sample space to consist of only the elementary events in A. All other possibilities have been eliminated. Now since A is assumed to be a proper subset of S, the original sample space has been reduced. Hence, if in using A as our *new* sample space we were to use the probabilities originally assigned to the elementary events in A, axiom (2) would be violated since the sum of these probabilities would

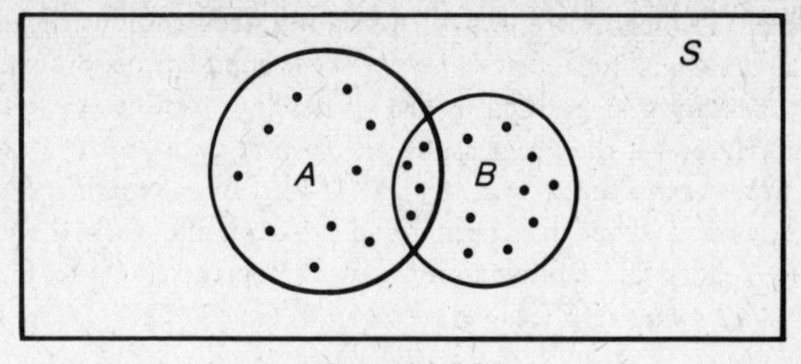

FIGURE 4.2

not equal 1. Therefore, we multiply each of the original probabilities of the elementary events in A by the same factor in such a manner that the sum of the adjusted probabilities is 1. Since the sum of the original probabilities was $P(A)$, it is clear that the desired multiplying factor is $1/P(A)$.

In multiplying each of the original probabilities by the same factor, we are preserving their relative magnitudes. For example, if the probability of a particular elementary event was twice the probability of another elementary event under the original assignment, it still is so under the modified assignment.

Now the only way in which both A and B can occur is for one of the elementary events in $A \cap B$ to occur. But if each of these probabilities is multiplied by $1/P(A)$ and the products are added together, we obtain $P(A \cap B)/P(A)$ for the desired probability. We therefore define the conditional probability of B given A as follows:

$$P(B|A) = \frac{P(A \cap B)}{P(A)} \qquad P(A) \neq 0 \tag{1}$$

It can be shown that the function defined above is a probability function; i.e., it satisfies the three axioms used to define a probability function.

If we multiply both sides of Equation (1) by $P(A)$ we obtain the following.

Theorem VII. $P(A \cap B) = P(A) \cdot P(B|A)$

This is the **multiplication law of probability.** It can be extended to cases involving the intersection of more than two events. For example,

$$P(A \cap B \cap C) = P(A) \cdot P(B|A) \cdot P(C|A \cap B) \tag{2}$$

Example 3 Two cards are drawn from a full deck without replacement; i.e., the first card is not replaced before the second card is drawn. What is the probability that both cards are spades?

Solution Here A is the event *spade on the first draw* and B is the event *spade on the second draw.* Clearly $P(A) = \frac{1}{4}$. In calculating $P(B|A)$ we note that after 1 spade has been removed from the deck there are 12 remaining spades and 51 remaining cards altogether. Hence $P(B|A) = 12/51 = 4/17$. Hence $P(A \cap B) = 1/4 \cdot 4/17 = 1/17$.

Example 4 Two cards are drawn from a full deck without replacement. What is the probability that they will consist of 1 heart and 1 spade?

Solution Note that in this case the specified event can occur in two different orders, i.e., heart, spade and spade, heart and that these are mutually exclusive events. Using the same line of reasoning employed in the previous example, we can easily see that the probability of drawing first a heart and then a spade is $1/4 \cdot 13/51 = 13/204$, and it is clear that the probability of drawing the cards in reverse order is the same. Hence the desired probability is $13/204 + 13/204 = 13/102$. Note that in this example we have used both the addition and multiplication laws of probability.

Example 5 Three cards are drawn from a full deck without replacement. What is the probability that all three are spades?

Solution Using the same line of reasoning employed in Example 4, we find the required probability is $1/4 \cdot 4/17 \cdot 11/50 = 11/850$.

4.6 Independent Events

Two events A and B are said to be independent when the occurrence of one has no effect on the probability of the occurrence of the other. More precisely, an event B is independent of an event A if and only if $P(B|A) = P(B)$. It is easily shown that if neither $P(A)$ nor $P(B)$ is 0, $P(B|A) = P(B)$ implies $P(A|B) = P(A)$. Hence if B is independent of A, then A is independent of B.

An immediate consequence of Theorem VII and the foregoing definition of independence is the following.

Theorem VIII. If A and B are independent events, then

$$P(A \cap B) = P(A) \cdot P(B)$$

The notion of independence can be extended to sets of more than two events. If we designate a set of n independent events by A_1, $A_2, \ldots, A_n$, then

$$P(A_1 \cap A_2 \cdots \cap A_n) = P(A_1) \cdot P(A_2) \cdot \cdots \cdot P(A_n)$$

Although our discussion of independence has proceeded under the assumption that $P(A) \neq 0$ and $P(B) \neq 0$, Theorem VIII can be taken to hold in general if we enlarge the definition of independence by saying that any two events are independent if at least one of them has probability 0.

Example 1 Three cards are drawn from a full deck with replacement; i.e., after each card is drawn it is replaced and the deck is thoroughly shuffled before the next card is drawn. What is the probability that all three cards are hearts?

Solution Here the three events are clearly independent and, by the formula shown above, the desired probability is evidently $1/4 \cdot 1/4 \cdot 1/4 = 1/64$.

4.7 Additional Examples

The following examples will serve to illustrate several of the previously stated formulas.

Example 1 An urn contains 6 red and 4 blue balls. Two balls are drawn without replacement. Find the probability of each of the following events:

(a) Both balls are red.

(b) One ball is red and one is blue.

(c) At least one ball is blue.

(d) Both balls are of the same color.

Solution

(a) $P(\text{both balls are red}) = \frac{6}{10} \cdot \frac{5}{9} = \frac{1}{3}$

(b) P (one is red and one is blue)

$$= P(\text{red, blue}) + P(\text{blue, red})$$

$$= \frac{6}{10} \cdot \frac{4}{9} + \frac{4}{10} \cdot \frac{6}{9} = \frac{8}{15}$$

(c) The events *both balls are red* and *at least one ball is blue* are complementary events. Hence

$$P(\text{at least one ball is blue}) = 1 - P(\text{both balls are red})$$

$$= 1 - \frac{1}{3} = \frac{2}{3}$$

(d) P (both balls are of the same color)

$$= P\text{ (both balls are red)} + P\text{ (both balls are blue)}$$

$$= \frac{1}{3} + \frac{4}{10} \cdot \frac{3}{9} = \frac{1}{3} + \frac{2}{15} = \frac{7}{15}$$

The permutation and combination formulas discussed in Chapter 1 can often be used to advantage in solving probability problems. Here are some examples:

Example 2 Part (a) of Example 1 may be alternatively solved as follows. The total number of ways of drawing 2 balls without replacement from an urn containing 10 balls is $\binom{10}{2}$. The number of ways of drawing 2 red balls is $\binom{6}{2}$. Hence

$$P\text{ (both balls are red)} = \frac{\binom{6}{2}}{\binom{10}{2}}$$

$$= \frac{6!}{2!4!} \cdot \frac{2!8!}{10!} = \frac{1}{3}$$

This agrees with the result previously obtained by the multiplication law.

Example 3 Part (d) of Example 1 may be solved as follows. The number of ways of drawing 2 red balls or 2 blue balls is $\binom{6}{2} + \binom{4}{2}$. Hence

$$P\text{ (both balls are of the same color)} = \frac{\binom{6}{2} + \binom{4}{2}}{\binom{10}{2}}$$

$$= \frac{21}{45} = \frac{7}{15}$$

the result previously obtained.

The permutation formulas of the previous chapter may also be used to advantage in certain probability problems. Here are two examples.

Example 4 Five persons file into a room in random order. What is the probability that a specified person will be first in line?

Solution The total number of orders in which the 5 persons can file into the room is 5! If a specified person is placed first in the line, then there are 4! ways in which the other 4 persons can be arranged in the file. Therefore the required probability is $4!/5! = 1/5$. This result is, of course, intuitively obvious since it is equally likely that each of the 5 persons will be first in line.

Example 5 Six books are arranged on a shelf in random order. What is the probability that 2 specified books will be next to each other?
Solution The 6 books can be arranged on the shelf in 6! different orders. If we think of the 2 specified books that are to be next to each other as 1 book, then there are only 5 objects to arrange, and these can be arranged in 5! different orders. However, the 2 specified books can be placed on the shelf next to each other in 2 different orders. Hence the total number of arrangements with the 2 books next to each other is $2 \cdot 5!$ Therefore the required probability is $2 \cdot 5!/6! = 1/3$.

4.8 Random Variables and Probability Distributions

We now continue our study of elementary probability with a discussion of random variables and the associated notion of a probability distribution. A **random variable** is a variable that is associated with a sample space in such a way that to each element of the sample space there corresponds a single value of the random variable.

It is customary to distinguish between a random variable and its possible values by using a capital letter to represent the random variable and the corresponding small letter, sometimes with subscripts, to represent its values. Thus a random variable might be represented by X and its three possible values by x_1, x_2, and x_3.

For an example of a random variable, consider the experiment of tossing a coin twice, and define the random variable X to be the number of heads appearing. Evidently in this case the possible values of X are 0, 1, and 2. With reference to the sample space, $S = \{HH, HT, TH, TT\}$, the value 0 corresponds to the outcome TT, the value 1 corresponds to the outcomes HT and TH, and the value 2 corresponds to the outcome HH.

A **probability distribution** is simply a correspondence between the possible values of a random variable and their associated probabilities. For example, let us determine the probability distribution of the random variable X, defined in the previous paragraph. If the coin is balanced, the probability of each of the outcomes in S is $\frac{1}{4}$. Hence the probability distribution of X is as shown in the table below.

x	0	1	2
$p(x)$	$\frac{1}{4}$	$\frac{1}{2}$	$\frac{1}{4}$

A graphical representation of this same probability distribution is shown in Figure 4.3. It consists of a series of line segments, the length of each proportional to the probability corresponding to the value of x shown at the base at the segment.

Probability distributions may also be defined by algebraic expressions. Examples of this type of representation will be shown presently.

Furthermore, it is clear that to qualify as a probability distribution, a function $p(x)$ must possess the following properties:

(1) $p(x) \geq 0$, for all values of x.

(2) $\sum_{\text{all } x} p(x) = 1$

Property (1) follows from axiom (1), and property (2) follows from axioms (2) and (3) in the definition of a probability function.

A word may be in order with regard to the distinction between a probability function that has been defined axiomatically and a function of the type that we have called a probability distribution. The domain of a probability function is a set of sets, i.e., events. The domain of a probability distribution is a set of real numbers

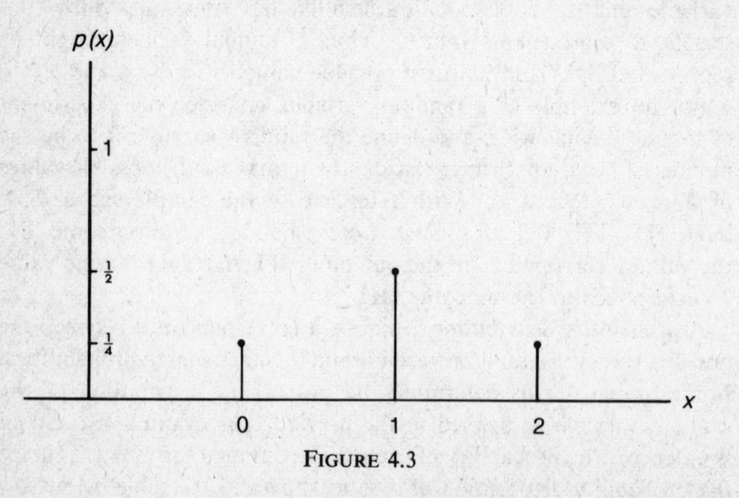

FIGURE 4.3

that are the possible values of a random variable. The range in both cases is a subset of the real numbers in the interval [0,1].

4.9 The Binomial and Hypergeometric Distributions

Of the probability distributions that may be represented by algebraic expressions, two of the best known, and most widely used are the binomial and the hypergeometric, which will now be discussed.

The **binomial distribution** will be introduced by means of a problem. Five cards are drawn from a full deck with replacement. What is the probability that exactly 3 of these cards will be hearts?

It is clear that the specified result can be obtained in a number of different orders. For example, HHHNN represents the order in which the first 3 cards drawn are hearts and the last 2 are nonhearts. The probability of this outcome is $\frac{1}{4} \cdot \frac{1}{4} \cdot \frac{1}{4} \cdot \frac{3}{4} \cdot \frac{3}{4} = (\frac{1}{4})^3 (\frac{3}{4})^2$. It is clear that any of the other possible orders in which exactly 3 hearts will occur (for example, NHNHH) will have the same probability. Therefore the total probability will be obtained by multiplying $(\frac{1}{4})^3 (\frac{3}{4})^2$ by the number of different orders in which exactly 3 hearts can occur. This can be obtained by noting that in choosing a particular order we are simply selecting the 3 positions out of 5 at which the hearts will appear. This can be done in $\begin{bmatrix} 5 \\ 3 \end{bmatrix}$ ways. Therefore the desired probability is

$$\binom{5}{3}\left(\frac{1}{4}\right)^3\left(\frac{3}{4}\right)^2 = \frac{5!}{3!2!}\,\frac{1}{64}\,\frac{9}{16} = \frac{45}{512}$$

The procedure employed in the previous example can be generalized as follows. If an experiment consists of n independent trials, in which the outcome of each trial is called either a success or a failure and if the probability of success on each trial is a constant denoted by p, then the probability $p(x)$ of exactly x successes is given by

$$p(x) = \binom{n}{x} p^x (1-p)^{n-x} \qquad x = 0, 1, \cdots, n \tag{1}$$

This is known as the **binomial distribution,** which has many applications.

Example 1 Assuming that the probability of a male birth is $\frac{1}{2}$, what is the probability that a family of 4 children will consist of 2 boys and 2 girls?

Solution Here $n=4$, $p=\frac{1}{2}$, and $1-p=\frac{1}{2}$. Hence the desired probability is

$$\binom{4}{2}\left(\frac{1}{2}\right)^2\left(\frac{1}{2}\right)^2=\binom{4}{2}\left(\frac{1}{2}\right)^4=\frac{3}{8}$$

Example 2 If in Example (1) x represents the number of boys in the family, obtain the probability distribution for X in algebraic, tabular, and graphical form.

Solution Using the formula for the binomial distribution with $n = 4$, $p = \frac{1}{2}$, we have

$$p(x)=\binom{4}{x}\left(\frac{1}{2}\right)^x\left(\frac{1}{2}\right)^{4-x}$$

$$=\binom{4}{x}\left(\frac{1}{2}\right)^4 \qquad x=0,1,2,3,4$$

Using the above expression for $p(x)$, we can readily calculate the values shown in the table below.

x	0	1	2	3	4
$p(x)$	$\frac{1}{16}$	$\frac{1}{4}$	$\frac{3}{8}$	$\frac{1}{4}$	$\frac{1}{16}$

In Figure 4.4 we have represented this distribution graphically by means of a **histogram,** a series of rectangles, centered above the possible values of x, whose heights are proportional to the corresponding probabilities.

We note that since the length of the base of each rectangle is 1, its area is numerically equal to the corresponding probability, and the total

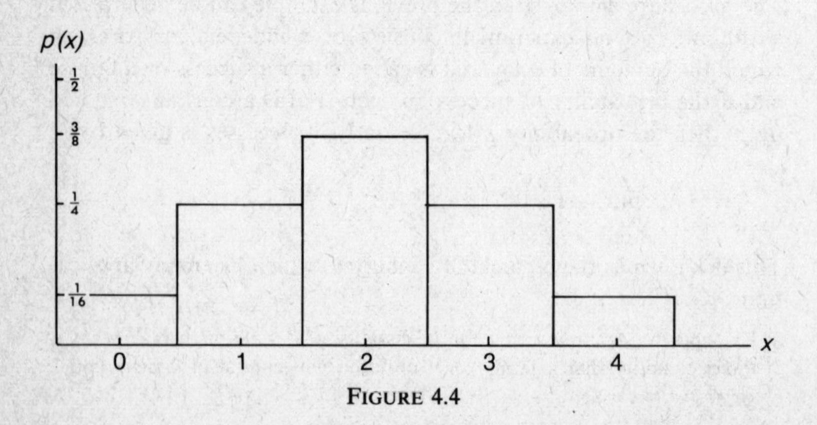

FIGURE 4.4

area of this histogram is 1. The association between probability and area is an important one and we shall return to it in a subsequent chapter.

Example 3 A sharpshooter hits the target $\frac{5}{6}$ of the time. If she takes 5 shots, what is the probability that she will score at least 3 hits?
Solution:

$$\begin{aligned} P\,(\text{at least 3 hits}) &= P\,(\text{exactly 3 hits}) + P\,(\text{exactly 4 hits}) \\ &\quad + P\,(\text{exactly 5 hits}) \\ &= \binom{5}{3}\left(\frac{5}{6}\right)^3\left(\frac{1}{6}\right)^2 + \binom{5}{4}\left(\frac{5}{6}\right)^4\left(\frac{1}{6}\right) + \left(\frac{5}{6}\right)^5 \\ &= \frac{1875}{1944} \simeq 0.96 \end{aligned}$$

Next we shall discuss the **hypergeometric distribution;** again we shall start with a problem. Suppose that from a group of 6 men and 4 women a committee of 4 is to be chosen by lot. What is the probability that the committee will consist of 2 men and 2 women?

Evidently a committee of 4 persons can be drawn from a group of 10 persons in $\binom{10}{4}$ different ways. (Note that we are assuming here that the drawing is being carried out without replacement.) The 2 men on the committee can be drawn from the 6 available men in $\binom{6}{2}$ ways, and the 2 women can be drawn from the 4 available women in $\binom{4}{2}$ ways. Hence the number of ways in which 2 men and 2 women can be drawn is the product of these two numbers, i.e., $\binom{6}{2}\cdot\binom{4}{2}$. Therefore the desired probability is

$$\frac{\binom{6}{2}\binom{4}{2}}{\binom{10}{4}} = \frac{\left(\frac{6!}{2!4!}\right)\left(\frac{4!}{2!2!}\right)}{\frac{10!}{4!6!}} = \frac{3}{7}$$

The procedure used above can be generalized as follows. Suppose that n objects are to be drawn without replacement from a total of N objects. Suppose furthermore that each object can be placed in one of two categories, let us say defective and nondefective, and that a of the N objects are defective. We wish to find the probability that exactly x of the n objects drawn will be defective. Using exactly the same reasoning employed in the previous numerical example, we see that the desired probability $p(x)$ is given by

$$p(x) = \frac{\binom{a}{x}\binom{N-a}{n-x}}{\binom{N}{n}}$$

This is known as the **hypergeometric distribution.** The domain of this function is a set of integers that depends on N, n, and a.

Example 4 A group of 12 manufactured items contains 4 defectives. If 5 items are drawn at random without replacement, what is the probability that exactly 2 of them will be defective?

Solution Here $N = 12$, $n = 5$, and $x = 2$. Hence the desired probability is

$$\frac{\binom{4}{2}\binom{8}{3}}{\binom{12}{5}} = \frac{\left(\frac{4!}{2!2!}\right)\left(\frac{8!}{3!5!}\right)}{\left(\frac{12!}{5!7!}\right)} = \frac{14}{33}$$

It may be noted that in problems involving drawing a number of items from a larger group, the binomial distribution applies if the drawing is carried out with replacement, whereas the hypergeometric distribution applies if it is done without replacement. However, in cases where the number of items drawn is small compared to the total number in the group, the binomial is often used to approximate the hypergeometric. The reason for this can be made clear through the following example.

Suppose a selection of 5 items is to be made from a group of 2000, of which 40 are defective. If the selection is made without replacement, then the probability of drawing a defective item will change on each draw so that, strictly speaking, the binomial distribution does not apply. However, the changes will be so small that the error resulting from the use of the binomial distribution will not be large. For example, the probability of obtaining a defective item on the first draw is $40/2000 = 0.02$. If a nondefective item is drawn, then the probability of obtaining a defective item on the second draw is $40/1999 \doteq 0.02001$. If a defective item is selected on the first draw, the probability of a defective on the second draw is $39/1999 \simeq 0.01951$. The changes in the probabilities for the other draws will be correspondingly small. Hence the use of the binomial distribution in a problem involving this experiment would not produce an error of serious proportions.

4.10 The Expected Value and the Variance of a Random Variable

We shall now discuss two important quantities associated with a random variable: the expected value and the variance.

The **expected value** of a random variable X, with probability distribution $p(x)$, is denoted by the symbol $E(X)$, and it is defined as follows:

$$E(X) = \sum_{\text{all } x} xp(x)$$

It is thus seen that the expected value of a random variable is obtained by multiplying each of its possible values by the corresponding probability and adding these products. For example, consider the following probability distribution:

x	1	2	3
$p(x)$	$\frac{1}{6}$	$\frac{1}{3}$	$\frac{1}{2}$

In this case,

$$E(X) = (1)\left(\tfrac{1}{6}\right) + (2)\left(\tfrac{1}{3}\right) + (3)\left(\tfrac{1}{2}\right) = \tfrac{7}{3}$$

Note that the expected value of X is not one of its possible values. Clearly, the word "expected" is used in this connection in a sense that differs from its ordinary meaning. Actually the value we have calculated is what is called a weighted average of the possible values of X. The weights are the corresponding probabilities. For this reason the expected value of a random variable is also referred to as its mean and is often denoted by the Greek letter μ.

In order to interpret further the value we have obtained for $E(X)$ let us consider an experiment that corresponds to the above probability distribution. Suppose an urn contains 6 numbered chips, of which 1 has the number 1, while 2 have the number 2, and 3 have the number 3. The experiment consists of drawing 1 chip at random from the urn. The random variable X is defined as the number of the chip that is drawn. Clearly, the probabilities will be as shown in the above table. Then the number $\frac{7}{3}$ that we have obtained as the expected value, may be interpreted as the long-run average value of X in the sense that if the experiment is repeated a large number of times, then the values of X can be expected to have an average very close to $\frac{7}{3}$.

The concept of expected value has many applications. One of these involves games of chance that are played for money or other objects of value. In this connection a fair game may be defined as one in which the expected value of each player's winnings is 0. This seems to be a reasonable definition since under these conditions the players may expect to break even in the long run.

Example 1 A game is played by two persons, A and B, as follows. A throws two dice. If the sum is 7, she wins \$3 from B; if the sum is 8, she loses \$2; and if the sum is 3, she loses \$4. Otherwise no money changes hands. Is this a fair game?

Solution Let X be a random variable representing A's winnings. Then the probability distribution for X is as shown in the following table. The probabilities can easily be verified by the student. Note that the losses are indicated by negative values of X.

x	0	3	-2	-4
$p(x)$	$\frac{23}{36}$	$\frac{6}{36}$	$\frac{5}{36}$	$\frac{2}{36}$

Then $$E(X) = (0)\left(\frac{23}{36}\right) + (3)\left(\frac{6}{36}\right) + (-2)\left(\frac{5}{36}\right) + (-4)\left(\frac{2}{36}\right) = 0$$

Since there are only two players in the game, it is evident that the expected value of B's winnings is also 0. Hence the game is fair.

It might be noted that most games such as roulette that are played in gambling casinos are not fair games in the sense defined above since the expected value of a customer's winnings is negative.

The application of the fair game concept to gambling games may appear to be a trivial one, but the same principle has application in much more important areas of life, specifically, in the insurance business. The agreement between an insurance company and the insured as evidenced by the policy may be thought of as a game, and the fair game principle is applied in computing the premium.

For example, suppose that a policy insures an individual for \$1000 against the occurrence of an event that has a probability of occurrence of 1/20. What is the fair premium for the policy?

We denote the policy holder's gain by X and the premium by Y. Clearly X has only two possible values. If the event in question should occur, then the policy holder's net gain will be \$1000 minus the premium Y that has been paid. If the event should fail to occur, then the policy holder will have a loss equal to Y. Hence the probability distribution for X will be as shown below.

x	$1000-Y$	$-Y$
$p(x)$	$\frac{1}{20}$	$\frac{19}{20}$

Hence, applying the fair game principle yields

$$E(X) = (1000 - Y)\left(\frac{1}{20}\right) + (-Y)\left(\frac{19}{20}\right) = 0$$

Solving this equation for Y, we obtain

$$50 - \frac{1}{20}Y - \frac{19}{20}Y = 0$$

or

$$Y = \$50$$

It should be noted that in practice an additional amount would have to be added to the premium to cover administrative costs and to ensure a reasonable profit for the company.

The notion of the expected value of a random variable may be generalized to define the expected value of a function of that variable. If $f(X)$ is a function of a random variable X, then

$$E[f(X)] = \sum_{\text{all } x} f(x)p(x)$$

For example, consider the random variable X, whose probability distribution is as follows:

x	1	2
$p(x)$	$\frac{1}{4}$	$\frac{3}{4}$

We wish to obtain the expected value of $f(X) = X^2$.

Then $$E(X^2) = \sum x^2 p(x) = (1)(\tfrac{1}{4}) + (4)(\tfrac{3}{4}) = \tfrac{13}{4}$$

The above extension of the idea of expected value leads directly to the definition of the variance of a random variable X. The **variance,** denoted by var (X) or by the symbol σ^2, is defined as follows.

$$\text{var}(X) = \sigma^2 = E[(X-\mu)^2] = \sum_{\text{all } x} (x-\mu)^2\, p(x)$$

where, as previously noted, $\mu = E(X)$. To compute the variance, one obtains the square of the difference between each possible value of the variable and the mean, multiplies this difference by the corresponding probability, and then adds these products.

To illustrate, we return to the example we used to demonstrate the calculation of the mean. The probability distribution was as follows:

x	1	2	3
$p(x)$	$\frac{1}{6}$	$\frac{1}{3}$	$\frac{1}{2}$

We have previously obtained $\mu = E(X) = \frac{7}{3}$. Hence,

$$\sigma^2 = \sum_{\text{all } x} (x - \tfrac{7}{3})^2\, p(x) = (1-\tfrac{7}{3})^2(\tfrac{1}{6}) + (2-\tfrac{7}{3})^2(\tfrac{1}{3}) + (3-\tfrac{7}{3})^2(\tfrac{1}{2}) = \tfrac{5}{9}$$

Closely associated with the variance of a random variable is a quantity called the standard deviation. The **standard deviation,** denoted by the letter σ, is defined as the positive square root of the variance:

$$\sigma = \sqrt{\sum_{\text{all } x} (x-\mu)^2 p(x)}$$

For the example above we obtain $\sigma = \sqrt{5}/3$.

It has already been pointed out that the expected value of a random variable is a long-run average value. An interpretation of the variance may readily be obtained by studying its definition. The variance is actually a weighted mean of the squares of the differences between the values of the random variable and its mean, the weights being the probabilities. Evidently the variance will tend to be large when these differences are large. For this reason the variance may be thought of as a measure of variability or dispersion. The two distributions whose graphs are shown in Figure 4.5 illustrate this interpretation of the variance.

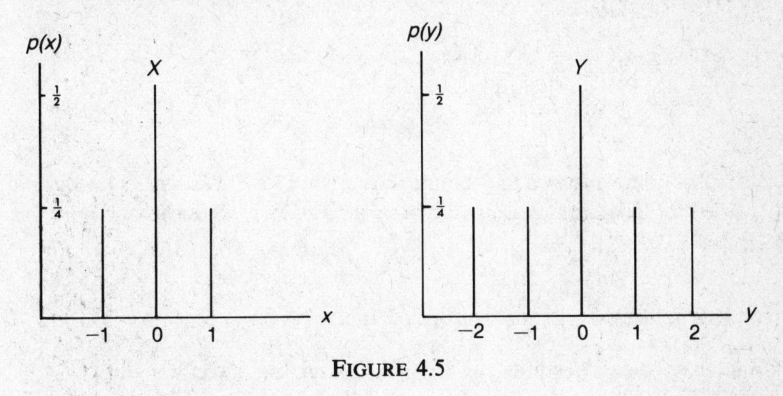

FIGURE 4.5

It is easily verified that $\mu_X = \mu_Y = 0$. However, $\sigma_X^2 = \frac{1}{2}$, whereas $\sigma_Y^2 = \frac{5}{4}$. The greater variability of Y as compared to X is reflected in its larger variance.

The formula previously given in defining the variance of a random variable can be transformed into a form which is preferable for computational purposes in many cases. This new formula is contained in the following theorem.

Theorem IX. If X is a random variable with probability distribution $p(x)$ and mean μ, then

$$\text{var}(X) = \sum_{\text{all } x} x^2 p(x) - \mu^2$$

Proof By definition

$$\text{var}(X) = \sum_{\text{all } x} (x - \mu)^2 p(x)$$

Noting that μ is a constant and using some of the properties of the summation symbol, we have

$$\sum(x-\mu)^2p(x) = \sum(x^2-2\mu x+\mu^2)p(x)$$

$$= \sum x^2p(x) - \sum 2\mu xp(x) + \sum \mu^2 p(x)$$

$$= \sum x^2p(x) - 2\mu \sum xp(x) + \mu^2 \sum p(x)$$

But $\Sigma\ xp(x) = \mu$ and $\Sigma\ p(x) = 1$

Hence $$\text{var}(X) = \sum x^2p(x) - 2\mu^2 + \mu^2$$

$$= \sum x^2p(x) - \mu^2$$

If this formula is used to compute the variance for the example previously used to illustrate the computation of the variance, we have

$$\sigma^2 = (1)(\tfrac{1}{6}) + (4)(\tfrac{1}{3}) + (9)(\tfrac{1}{2}) - \tfrac{49}{9} = \tfrac{5}{9}$$

This agrees with the previous calculation.

4.11 Special Formulas for the Mean and the Variance

In certain cases the mean and the variance of a random variable with a given probability distribution can be calculated much more efficiently using special formulas applicable to the given distribution. In such cases, we often refer to the mean and the variance of the distribution rather than of the random variable.

Let us consider a binomial distribution with $n = 4$ and $p = \frac{1}{2}$. By use of the binomial distribution formula, the following table is easily obtained:

x	0	1	2	3	4
$p(x)$	$\frac{1}{16}$	$\frac{4}{16}$	$\frac{6}{16}$	$\frac{4}{16}$	$\frac{1}{16}$

Hence $$\mu = \sum xp(x) = (0)(\tfrac{1}{16}) + (1)(\tfrac{4}{16}) + (2)(\tfrac{6}{16})$$

$$+ (3)(\tfrac{4}{16}) + (4)(\tfrac{1}{16})$$

$$= 2$$

The result is hardly unexpected since it indicates that for an experiment consisting of 4 trials with probability of success equal to $\frac{1}{2}$

on each trial, the average number of successes is 2. This result could evidently have been obtained by calculating the product np. Using the algebraic expression for $p(x)$, it can be shown that the formula just indicated holds in general. This, along with a general formula that can be derived for the variance of the binomial distribution, may be written as follows:

$$\mu = np$$
$$\sigma^2 = np(1 - p)$$

Hence, for the binomial distribution discussed above,

$$\sigma^2 = (4)(\tfrac{1}{2})(\tfrac{1}{2}) = 1$$

A general formula for the mean of a hypergeometric distribution is given by

$$\mu = n \cdot \frac{a}{N}$$

This is seen to be similar to the formula for the mean of a binomial distribution since it involves multiplying the number of draws by the probability of a defective item on the first draw. The formula for the variance of the hypergeometric distribution is considerably more complex and will be omitted.

For a final illustration of a special formula for the mean of a probability distribution we consider the following distribution:

$$p(x) = \frac{1}{n} \qquad x = 1, 2, \ldots, n$$

Since the probability is constant for all values of x, this may appropriately be called the **uniform** distribution for a discrete random variable. (There is also a uniform distribution for a continuous random variable.)

We may obtain the mean of the uniform distribution as follows:

$$E(x) = \sum_{\text{all } x}^{n} xp(x) = \frac{1}{n} \sum_{\text{all } x} x$$

But $\sum_{\text{all } x} x$ is simply the sum of the integers from 1 to n, and by a well-known formula it equals $n(n + 1)/2$.

Hence $$E(X) = \frac{1}{n} \cdot \frac{n(n + 1)}{2} = \frac{n + 1}{2}$$

4.12 Countably Infinite Sample Spaces

In Section 4.3 a probability function was defined by means of three axioms. However, axiom (3), as stated, applied only to finite sample spaces. For a countably infinite sample space, axiom (3) must be modified as follows. If $A_1, A_2, \ldots$ is an infinite sequence of mutually exclusive events, then

$$P(A_1 \cup A_2 \cup \cdots) = P(A_1) + P(A_2) + \cdots$$

It should of course be noted that when we refer to the sum of the terms on the right we are using the word "sum" in a special sense. Such a series is said to have a sum if and only if the sum of a finite number of terms $A_1 + A_2 + \cdots + A_n$ can be brought as close as we wish to some number by taking a sufficiently large value of n. This number is called the sum of the series.

For an illustration of the calculation of a probability involving a countably infinite sample space, we recall the experiment referred to in Section 4.2. The experiment consists of conducting a series of tosses of a coin until a head occurs, the outcome of the experiment being defined as the number of the toss on which the first head occurs. As previously stated, the sample space is $S = [1,2,\ldots]$.

Let us consider the problem of calculating the probability that the outcome of the experiment is an even number. We assume that the probability of a head on each toss is $\frac{1}{2}$, and we denote the outcome of the experiment by x. Then using H and T to denote head and tail and using the multiplication law for independent events, we can construct the following table.

SEQUENCE	X	PROBABILITY
H	1	$\frac{1}{2}$
TH	2	$\frac{1}{4}$
TTH	3	$\frac{1}{8}$
TTTH	4	$\frac{1}{16}$
⋮	⋮	⋮

The required sum is the sum of a geometric series of the form $a + ar + ar^2 + \cdots$. It is shown in books on algebra that if $|r| < 1$, the sum of this series is $a/(1-r)$. Hence the required probability is $P(x = 2, 4, 6, \cdots) = \frac{1}{4} + \frac{1}{16} + \frac{1}{64} + \cdots = \frac{1}{4}/\frac{3}{4} = \frac{1}{3}$.

Using the same procedure, the student may readily verify that $P(S) = 1$, i.e., that axiom (2) is satisfied for this sample space.

4.13 Bayes' Theorem

Up to this point we have been concerned with the problem of determining the probability of a specified event *before* the performance of a given experiment. We now discuss a procedure for calculating the probability of an event associated with the experiment *after* the experiment has been performed.

To introduce this idea, we consider the following experiment involving two urns. Urn 1 contains 4 red and 2 blue balls. Urn 2 contains 2 red and 6 blue balls. An urn is to be chosen at random and a ball is to be drawn from the urn. We shall designate certain events as follows.

A—Urn 1 is chosen.

A'—Urn 2 is chosen.

B—A red ball is drawn.

Note that the events designated as A and A' are in fact complementary. First consider the problem of determining the probability that a red ball will be drawn. Assuming $P(A) = P(A') = \frac{1}{2}$, we have

$$\begin{aligned} P(B) &= P(A) \cdot P(B|A) + P(A') \cdot P(B|A') \\ &= (\tfrac{1}{2})(\tfrac{2}{3}) + (\tfrac{1}{2})(\tfrac{1}{4}) = \tfrac{11}{24} \end{aligned}$$

The above problem evidently involves nothing that has not been previously discussed. Let us now consider a second and somewhat different problem. Suppose that the experiment has already been performed and a red ball has been drawn. What is the probability that urn 1 was chosen, i.e., what is $P(A|B)$? We have

$$P(A\,|\,B) = \frac{P(A \cap B)}{P(B)} = \frac{P(A)P(B|A)}{P(A)P(B|A) + P(A')P(B|A')}$$

where we have replaced $P(B)$ by the expression used in the previous problem. Hence

$$P(A|B) = \frac{(\frac{1}{2})(\frac{2}{3})}{(\frac{1}{2})(\frac{2}{3}) + (\frac{1}{2})(\frac{1}{4})} = \frac{\frac{1}{3}}{\frac{11}{24}} = \frac{8}{11}$$

Note that prior to the experiment the probability of drawing urn 1 was $\frac{1}{2}$. But our knowledge of the outcome of the experiment has produced a modified value for this probability. Intuitively we might have expected that the modified value would be greater than the original value, since the proportion of red balls in urn 1 is greater than in urn 2.

The probability formula used in the above problem is a special case of what is known as **Bayes' theorem.** In its more general form it may be stated as follows:

Theorem X. If $A_1, A_2, \ldots, A_n$ are mutually exclusive events such that $S = A_1 \cup A_2 \cup \cdots \cup A_n$, then

$$P(A_i|B) = \frac{P(A_i)P(B|A_i)}{P(A_1)P(B|A_1) + P(A_2)P(B|A_2) + \cdots + P(A_n)P(B|A_n)}$$

$$i = 1, 2, \ldots, n$$

The proof is similar to the one used in connection with the special case.

For a second illustration of the use of Bayes' theorem, we consider the following problem. A manufacturer buys a certain part from three companies C_1, C_2, and C_3. The percentages of these parts that are bought from these companies are 40, 35, and 25, respectively. Furthermore, the percentages of defective parts that are obtained from these companies are 1, 2, and 3, respectively. If a part is found to be defective, what is the probability that it was bought from C_1?

We may designate certain events as follows:

A_1—The part was bought from C_1.

A_2—The part was bought from C_2.

A_3—The part was bought from C_3.

B—The part is defective.

Then

$$P(A_1|B) = \frac{(0.40)(0.01)}{(0.40)(0.01) + (0.35)(0.02) + (0.25)(0.03)} \simeq 0.22$$

4.14 A Surprising Example

We shall close this chapter with an example interesting in that the result is one that seems to defy our intuitive logic. It concerns

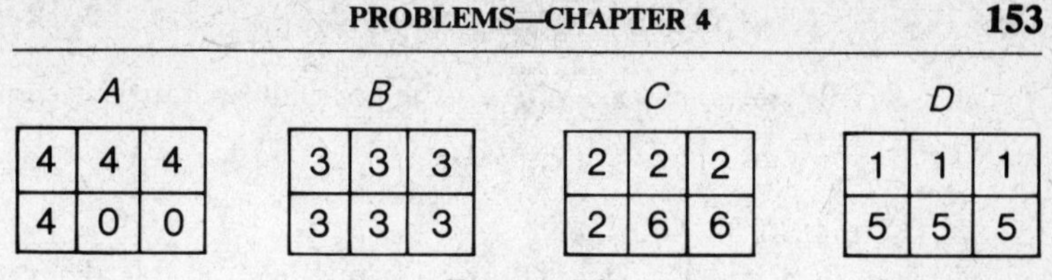

FIGURE 4.6

a set of unconventional dice which have been devised by Professor Bradley Efron of Stanford University. The set of four dice is represented in Figure 4.6 and the six numbers represent the numbers of spots appearing on each face of the dice.

A simple game is to be played with these dice. One player is allowed to select one of the dice. An opponent may then select any one of the remaining three dice. Each player then throws his or her die, and the player obtaining the higher number wins the game. Now, it is easily seen that the probability that the player holding die *A* will beat the player holding die *B* is $\frac{2}{3}$. This is true because of the 36 possible combinations involving die *A* and die *B* 24 of them consist of 4 on *A* and 3 on *B*, whereas the other 12 consist of 0 on *A* and 3 on *B*. In the same way it can readily be seen that the probability that *B* will beat *C* is also $\frac{2}{3}$ and that *C* has the same probability of beating *D*.

Now comes the big surprise: If you study the various possible combinations of *D* and *A*, you will find that the probability is $\frac{2}{3}$ that *D* will beat *A!*

The surprising thing is that these dice do not obey what is known as the transitive law. A well-known relationship that obeys the transitive law is the relationship, *is greater than.* For example, if $a > b$ and $b > c$, then it follows that $a > c$. One would expect the same property to hold here, but it does not.

Furthermore, note that the player who is allowed to select a die first has a distinct disadvantage. His or her opponent can always select a die that gives a $\frac{2}{3}$ probability of winning.

Problems—Chapter 4

1. Given the sample space $S = \{a, b, c, d, e\}$, the probabilities $P\{a\} = P\{b\} = P\{c\} = \frac{1}{5}$, $P\{d\} = \frac{1}{4}$, and that $A = \{a, c, e\}$, find $P(A)$.
2. A single card is drawn from a full deck. Find the probability that the card is
   (a) A 7
   (b) A face card (ace, king, queen, or jack)
   (c) A heart or a face card

3. Two dice are thrown. What is the probability that their sum is
   (a) 11
   (b) at least 8
4. Two dice are thrown. Let x be a random variable defined as the sum of the two dice. Write out the entire probability distribution, i.e., make a table of x and $p(x)$.
5. An urn contains 8 red, 4 blue, and 3 white balls. Two balls are drawn without replacement. Find the probability that
   (a) Both balls will be red
   (b) One ball will be red and one will be blue
   (c) At least one ball will be red
   (d) The balls will be of the same color
   (e) The balls will not be of the same color
6. The probabilities that a student will pass math, history, and English are $\frac{1}{2}$, $\frac{2}{3}$, and $\frac{3}{4}$, respectively. Find the probability that she will
   (a) Pass all three subjects
   (b) Fail all three subjects [note that (a) and (b) are not complementary events]
   (c) Pass at least one of the subjects
   (d) Pass exactly two of the subjects
7. A lot contains 15 items, of which 4 are defective. If 3 items are drawn without replacement, what is the probability that they will include
   (a) Exactly 1 defective item
   (b) At least 1 defective item
8. A baseball player has a batting average of $0.333 \simeq \frac{1}{3}$. If he comes to bat 6 times, what is the probability that he will make
   (a) Exactly 2 hits
   (b) At most 2 hits
9. What is the probability that a family of 6 children will contain
   (a) 3 boys and 3 girls
   (b) At least 4 girls
10. Eight books are placed on a shelf in random order. What is the probability that 3 specified books will appear consecutively on the shelf?
11. Find the expected value and the variance of the following probability distribution:

x	-1	1	2
$p(x)$	$\frac{1}{4}$	$\frac{1}{2}$	$\frac{1}{4}$

12. Consider the following two-person game. *A* draws a single card from a full deck. If it is a heart, she wins \$12 from *B*. If it is a diamond, she loses \$8. If it is a black ace, she loses \$26. Otherwise no money changes hands. Is this a fair game?
13. Find the variance of the uniform distribution:

$$p(x) = \frac{1}{n} \qquad x = 1, 2, \ldots, n$$

$$\left(\text{Note:} \quad \sum_{x=1}^{n} x^2 = \frac{n(n+1)(2n+1)}{6}\right)$$

14. Urn 1 contains 5 red and 3 blue balls. Urn 2 contains 4 red and 2 blue balls. Urn 3 contains 3 red and 6 blue balls. An urn is chosen at random, and a ball is drawn from it. The ball is observed to be red. What is the probability that urn 1 was chosen?

5

ELEMENTARY CALCULUS

5.1 Introduction

The development of the branch of mathematics known as the **calculus** may be seen as an approach to the solution of two geometric problems. The first, the problem of finding the slope of the tangent to a curve at a specified point [Figure 5.1(a)] leads to the fundamental concept of what is called the **differential calculus.** The second, the problem of calculating the area of a region that is bounded at least in part by a given curve [Figure 5.1(b)], leads to the basic idea of the **integral calculus.**

However, the ramifications of this subject go far beyond the solution of geometric problems into a myriad of applications in both pure and applied mathematics. Thus, the calculus has become the central tool of mathematical analysis.

We shall now explore the basic concept of the differential calculus.

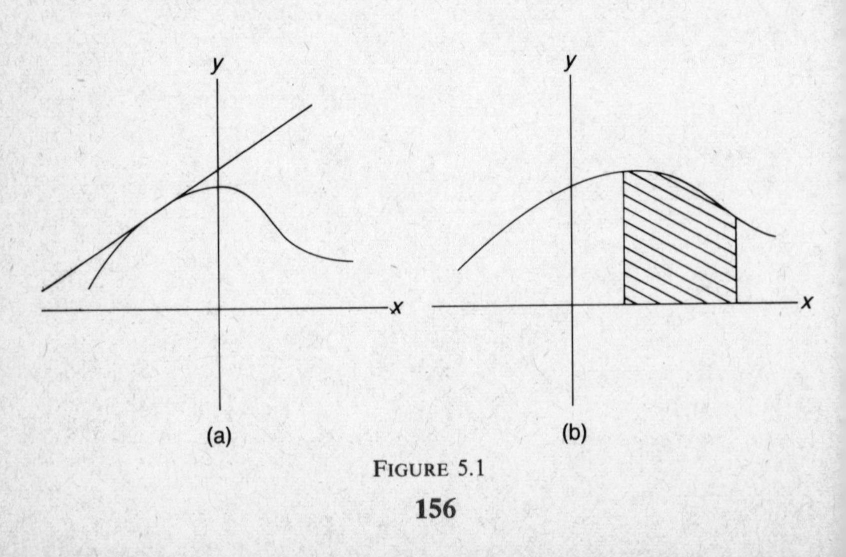

FIGURE 5.1

5.2 The Derivative

Consider the function $f(x) = x^2$. The difference between the values of the function at $x = 2$ and $x = 5$ is $f(5) - f(2) = 25 - 4 = 21$, and the change in the value of the function per unit change in x over the interval $2 \leq x \leq 5$ is

$$\frac{f(5) - f(2)}{5 - 2} = \frac{21}{3} = 7$$

We have thus calculated the average rate of change of the function over the specified interval. In general terms, if we designate the smaller value of x by x_1 and the larger value by x_2, then the average rate of change of $f(x)$ over the interval $x_1 \leq x \leq x_2$ is given by

$$\frac{f(x_2) - f(x_1)}{x_2 - x_1}$$

Alternatively, if we let $x_2 - x_1 = \Delta x$, then $x_2 = x_1 + \Delta x$, and the rate of change can be written

$$\frac{f(x_1 + \Delta x) - f(x_1)}{\Delta x}$$

Evidently Δx can be made as small as desired; the above expression is a rate of change over a very small interval. This suggests the possibility of obtaining the rate of change at a point by allowing Δx to approach 0. The limit, if it exists, which is approached by the average rate of change of the function as Δx approaches 0, is designated by the symbol $f'(x_1)$; therefore

$$f'(x_1) = \lim_{\Delta x \to 0} \frac{f(x_1 + \Delta x) - f(x_1)}{\Delta x}$$

or since x_1 may be any value of x in the domain of the function, we may write simply

$$f'(x) = \lim_{\Delta x \to 0} \frac{f(x + \Delta x) - f(x)}{\Delta x}$$

The function $f'(x)$ is called the **derivative of the function** $f(x)$. If the function is denoted by y instead of by $f(x)$, then the derivative is designated by the symbol y' or by dy/dx. In this last notation the symbol d/dx stands for the *operation* of calculating the derivative of y.

The process of obtaining the derivative of a function is called differentiation, and a function whose derivative exists at a point is said to be differentiable at that point. It can be shown that if a function is differentiable at a point, then it is continuous at that point. In other words, continuity is a *necessary* condition of differentiability. However, the converse is not true, since there exist functions that are continuous but that are not differentiable.

Returning to the function $f(x) = x^2$, we can obtain the derivative $f'(x)$ by the following steps:

$$f(x) = x^2$$

$$f(x+\Delta x) = (x+\Delta x)^2 = x^2 + 2x\,\Delta x + (\Delta x)^2$$

$$f(x+\Delta x) - f(x) = x^2 + 2x\,\Delta x + (\Delta x)^2 - x^2$$

$$= 2x\,\Delta x + (\Delta x)^2$$

$$\frac{f(x+\Delta x) - f(x)}{\Delta x} = 2x + \Delta x$$

$$f'(x) = \lim_{\Delta x \to 0} (2x + \Delta x) = 2x$$

Hence the derivative of the function x^2 at any point is $2x$. For example, $f'(2) = 4$; i.e., at the point $x = 2$ the function is increasing by 4 units per unit change in x. Since the derivative is the rate of change of a function, it is intuitively clear that a positive derivative indicates an increasing function whereas a negative derivative indicates one that is decreasing.

In the case of a function of the form $f(t)$, in which t represents time, Δt represents an interval of time. The expression

$$\frac{f(t+\Delta t) - f(t)}{\Delta t}$$

represents the average rate of change of the function during an interval of time. Furthermore, $f'(t)$ represents the rate of change at a particular moment of time, or as it is often called, the instantaneous rate of change.

For example, consider the function $s = t^3 - t$. We wish to find the instantaneous rate of change of s with respect to time.

$$s = t^3 - t$$

$$s + \Delta s = (t+\Delta t)^3 - (t+\Delta t)$$

$$= t^3 + 3t^2\,\Delta t + 3t(\Delta t)^2 + (\Delta t)^3 - t - \Delta t$$

$$\Delta s = t^3 + 3t^2\,\Delta t + 3t(\Delta t)^2 + (\Delta t)^3 - t - \Delta t - (t^3 - t)$$

$$= 3t^2\,\Delta t + 3t(\Delta t)^2 + (\Delta t)^3 - \Delta t$$

$$\frac{\Delta s}{\Delta t} = 3t^2 + 3t\,\Delta t + (\Delta t)^2 - 1$$

$$\frac{ds}{dt} = \lim_{\Delta t \to 0} [3t^2 + 3t\,\Delta t + (\Delta t)^2 - 1] = 3t^2 - 1$$

For another example, consider the situation referred to in Section 1.9, in which a quantity y is inversely proportional to a quantity x. This relationship is expressed by the following equation, where k is a constant:

$$y = \frac{k}{x}$$

$$y + \Delta y = \frac{k}{x + \Delta x}$$

$$\Delta y = \frac{k}{x + \Delta x} - \frac{k}{x} = k\left(\frac{1}{x + \Delta x} - \frac{1}{x}\right)$$

$$= k\left(\frac{x - (x + \Delta\,x)}{x(x + \Delta x)}\right) = \frac{-k\,\Delta x}{x(x + \Delta x)}$$

$$\frac{\Delta y}{\Delta x} = \frac{k}{x(x + \Delta x)}$$

$$\frac{dy}{dx} = \lim_{\Delta x \to 0}\left(\frac{-k}{x(x + \Delta x)}\right) = -\frac{k}{x^2}$$

Note that it has been shown that if y varies inversely with x, then the derivative dy/dx varies inversely with the square of x.

5.3 Geometric Interpretation of the Derivative

Consider the function $y = 3x - 5$. If we find the derivative dy/dx for this function by the process discussed in the previous section, we readily find $dy/dx = 3$. This value of the derivative is easily seen to have a simple geometric interpretation.

The student is familiar with the fact that the graph of the function $y = 3x - 5$ is a straight line. Furthermore, it has previously been

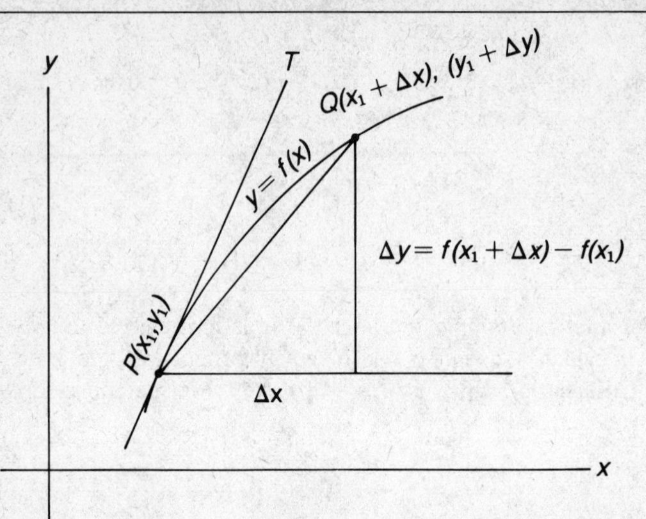

FIGURE 5.2

shown that when the equation of a line is written in the form $y = mx + b$, the value of m is the *slope* of the line, a measure of the steepness with which the line rises or falls as x increases. Hence the slope of the line discussed above is 3, which is also the value of dy/dx. It can be shown that the derivative of any linear function is the slope of the corresponding line. However, this geometric interpretation of the derivative can be extended to functions that are not linear, as will now be shown. The student will also note that this is the problem referred to in Section 5.1 that leads to the concept of the derivitive.

In Figure 5.2 is shown a portion of the graph of a differentiable function $y = f(x)$.

The points P and Q are two points on the curve whose x and y coordinates differ by Δx and Δy, respectively. Evidently, the ratio $\Delta y/\Delta x$ is the slope of the secant line PQ. It is clear that if Δx is allowed to approach 0, then Δy will also approach 0, and the point Q will move along the curve toward P. At the same time the secant line PQ will rotate and approach the position of the line PT, which is called the tangent line to the curve at the point P. Furthermore, the slope of the secant line PQ will approach the slope of the tangent line PT.

Since the slope of the secant line is given by $\Delta y/\Delta x$ and since the derivative dy/dx has been defined as

$$\frac{dy}{dx} = \lim_{\Delta x \to 0} \frac{\Delta y}{\Delta x}$$

the following geometric interpretation of the derivative may be stated.

If $y = f(x)$ is a differentiable function of x and if (x_1,y_1) is a point on the graph of the function, then $f'(x_1)$ is the slope of the tangent to the curve at the point (x_1,y_1).

Suppose, for example, that we desire to find the slope of the tangent to the curve $f(x) = x^2 - 3x$ at the point $(2,-2)$. We find $f'(x) = 2x - 3$. Hence the slope of the tangent line at the given point is $f'(2) = 4 - 3 = 1$.

5.4 Rules for Differentiation

The process by which the derivatives of various functions have been found can, if applied to general functions of various types, be used to derive simple rules for obtaining derivatives. Thereby we can reduce considerably the amount of time and effort involved.

For example, consider the function $f(x) = x^n$, where n is a positive integer. Applying our process for obtaining the derivative yields

$$f(x) = x^n$$
$$f(x + \Delta x) = (x + \Delta x)^n$$

Applying a well-known theorem of algebra known as the binomial theorem, we have

$$f(x + \Delta x) = x^n + nx^{n-1}\,\Delta x + \frac{n(n-1)x^{n-2}(\Delta x)^2}{2!} + \cdots + (\Delta x)^n$$

$$f(x + \Delta x) - f(x) = nx^{n-1}\,\Delta x + \frac{n(n-1)x^{n-2}(\Delta x)^2}{2!} + \cdots + (\Delta x)^n$$

$$\frac{f(x + \Delta x) - f(x)}{\Delta x} = nx^{n-1} + \frac{n(n-1)x^{n-2}(\Delta x)}{2!} + \cdots + (\Delta x)^{n-1}$$

$$f'(x) = \lim_{\Delta x \to 0} \left(nx^{n-1} + \frac{n(n-1)x^{n-2}(\Delta x)}{2!} + \cdots + (\Delta x)^{n-1}\right)$$
$$= nx^{n-1}$$

For example, if $f(x) = x^4$, then $f'(x) = 4x^3$; if $g(x) = x^5$, then $g'(x) = 5x^4$. Also if $f(x) = x$, $f'(x) = 1 \cdot x^{1-1} = 1 \cdot x^0 = 1 \cdot 1 = 1$. (Although x^0 is undefined when $x = 0$, this result is valid for all values of x.)

Furthermore, using the properties of limits contained in the theorems previously stated, we can establish the following properties of derivatives.

(1) If $y = kf(x)$, where $f(x)$ is a differentiable function and k is a constant, then $y' = kf'(x)$; i.e., the derivative of a constant times a function equals the constant times the derivative of the function.

(2) If $h(x) = f(x) + g(x)$, where $f(x)$ and $g(x)$ are differentiable functions, then $h'(x) = f'(x) + g'(x)$; i.e., the derivative of the sum of two functions equals the sum of their derivatives.

Applying these two properties to the function $y = 4x^3 - 7x^2 + 3x$, we have

$$\begin{aligned}\frac{dy}{dx} &= \frac{d}{dx}(4x^3) + \frac{d}{dx}(-7x^2) + \frac{d}{dx}(3x)\\ &= 4\frac{d}{dx}(x^3) - 7\frac{d}{dx}(x^2) + 3\frac{d}{dx}(x)\\ &= 4(3x^2) - 7(2x) + 3(1)\\ &= 12x^2 - 14x + 3\end{aligned}$$

The proof given for the formula for the derivative of $f(x) = x^n$ is valid only when n is a positive integer. However, it can be shown that the formula $f'(x) = nx^{n-1}$ holds when n is any real number. This important differentiation formula can therefore be used when n is a positive or a negative integer, a positive or a negative fraction, or even when n is an irrational number, i.e., a number such as $\sqrt{2}$ that cannot be expressed as the quotient of two integers.* In particular, the function $f(x) = k$, where k is a constant, may be written $f(x) = k \cdot x^0$. Its derivative is therefore $f'(x) = k \cdot 0x^{-1} = 0$. Hence the derivative of a constant is 0. This result might have been anticipated since the derivative has been defined as a rate of change and a constant function is one whose value does not change.

* However, x^n, where n is irrational, is defined only for $x > 0$.

Example 1:

$$f(x) = x^{1/2}$$

$$f'(x) = \frac{1}{2}x^{-1/2} = \frac{1}{2x^{1/2}}$$

Example 2:

$$f(x) = \frac{1}{x^2} = x^{-2}$$

$$f'(x) = -2x^{-3} = -\frac{2}{x^3}$$

Example 3:

$$f(x) = \frac{1}{x^{1/3}} = x^{-1/3}$$

$$f'(x) = -\frac{1}{3}x^{-4/3} = -\frac{1}{3x^{4/3}}$$

Example 4:

$$f(x) = 2x^{2/3} - 4x^{-3} + 5x^{\sqrt{2}} + 7$$

$$f'(x) = \frac{4}{3}x^{-1/3} + 12x^{-4} + 5\sqrt{2}\,x^{\sqrt{2}-1}$$

$$= \frac{4}{3x^{1/3}} + \frac{12}{x^4} + 5\sqrt{2}\,x^{\sqrt{2}-1}$$

5.5 The Chain Rule

Consider the function $y = (2x - 3)^2$. The rules for differentiation that have been previously stated do not cover a function of this form. However, if we expand the binomial expression, we obtain $y = 4x^2 - 12x + 9$; hence $y' = 8x - 12$. Note, however, that in the case of the function $y = \sqrt{3x - 5}$ we would not be able to proceed as in the previous example. In order to extend our rules of differentiation to cover functions of this type we need the important formula known as the **chain rule.**

Suppose y and u are related by the function $y = f(u)$, while u and x are related by the function $u = g(x)$. Then the chain rule states that the derivative of y with respect to x is equal to the derivative of y with respect to u multiplied by the derivative of u with respect to x. The truth of this rule is intuitively clear, involving as it does, the multiplication of rates. For example, if Ellen runs twice as fast as George, and Bill runs three times as fast as Ellen, then Bill runs six times as fast as George. In symbols the chain rule may be written

$$\frac{dy}{dx} = \frac{dy}{du}\frac{du}{dx}$$

It is assumed that y and u are differentiable functions. However, it should be emphasized that no meaning has been assigned to the symbols dx, dy, and du when standing alone, and hence the chain rule cannot be proved by simple algebraic cancellation in the above equation. A mathematically precise proof is given in most calculus textbooks.

We now return to our last two examples. For the function $y = (2x - 3)^2$, let $u = 2x - 3$. Then $y = u^2$ and $dy/du = 2u$. Also $du/dx = 2$. Hence, using the chain rule, we have

$$\frac{dy}{dx} = 2u \cdot 2 = 4u = 4(2x - 3) = 8x - 12$$

a result which agrees with the one previously obtained by another procedure. For the function $y = \sqrt{3x - 5}$ we let $u = 3x - 5$. Then

$$y = \sqrt{u} = u^{1/2}$$

Hence
$$\frac{dy}{du} = \frac{1}{2} u^{-1/2} = \frac{1}{2\sqrt{u}}$$

$$\frac{du}{dx} = 3$$

$$\frac{dy}{dx} = \frac{3}{2\sqrt{u}} = \frac{3}{2\sqrt{3x - 5}}$$

Using the chain rule, we can now generalize our previous formula for the derivative of the function $y = x^n$ as follows.

If $y = u^n$, where u is a differentiable function of x, then

$$\frac{dy}{dx} = nu^{n-1}\frac{du}{dx}$$

Using this formula, we can readily differentiate many functions without writing out the substitutions as we did in the previous examples.

For example, in differentiating the function $y = (x^2 - 3)^7$ we simply think of $x^2 - 3$ as u and apply the above formula to obtain

$$\frac{dy}{dx} = 7(x^2 - 3)^6 \cdot (2x) = 14x(x^2 - 3)^6$$

The composite function notation can be used to state the chain rule in another form. If f and g are differentiable functions, then

$$\frac{d}{dx}\{f[g(x)]\} = f'[g(x)] \cdot g'(x)$$

For example, consider the function $h(x) = \sqrt{x^2 - 5}$. Here we may think of $h(x)$ as the composite function $f[g(x)]$, where $f(x) = \sqrt{x}$ and $g(x) = x^2 - 5$. Hence $f'(x) = 1/2\sqrt{x}$ and $g'(x) = 2x$, so that

$$\begin{aligned} h'(x) &= \frac{d}{dx}\{f[g(x)]\} \\ &= f'[g(x)]\, g'(x) \\ &= \frac{1}{2\sqrt{x^2 - 5}}\, 2x \\ &= \frac{x}{\sqrt{x^2 - 5}} \end{aligned}$$

5.6 Higher-Order Derivatives

The derivative of a function of x is itself a function of x; assuming that it is differentiable, we can calculate its derivative. This is called the **second derivative** of the original function and it is denoted by the symbols $f''(x)$, y'', or d^2y/dx^2. If we differentiate again, we obtain the third derivative of the original function, denoted by $f'''(x)$, y''', or d^3y/dx^3. This process can be continued to obtain derivatives of higher order.

Example 1:

$$\begin{aligned} y &= 4x^3 - 3x^2 + 12x - 4 \\ y' &= 12x^2 - 6x + 12 \\ y'' &= 24x - 6 \\ y''' &= 24 \\ y^{\text{iv}} &= 0 \end{aligned}$$

Example 2:

$$y = (2x-5)^4$$
$$y' = 4(2x-5)^3 \cdot 2 = 8(2x-5)^3$$
$$y'' = 8 \cdot 3(2x-5)^2 \cdot 2 = 48(2x-5)^2$$
$$y''' = 48 \cdot 2(2x-5) \cdot 2 = 192(2x-5) = 384x - 960$$

Example 3:

$$y = (3x-2)^{1/2}$$
$$y' = \frac{1}{2}(3x-2)^{-1/2} \cdot 3 = \frac{3}{2(3x-2)^{1/2}}$$
$$y'' = \frac{3}{2} \cdot \left(-\frac{1}{2}\right)(3x-2)^{-3/2} \cdot 3 = -\frac{9}{4(3x-2)^{3/2}}$$

5.7 Derivatives of Products and Quotients

In some cases, in obtaining derivatives it is necessary or desirable to think of the function as the product of two other functions. For example, the function $f(x) = x^3\sqrt{x-5}$, can be thought of as the product of the two functions $g(x) = x^3$ and $h(x) = \sqrt{x-5}$. For situations of this kind the following theorem may be employed.

Theorem I. If $f(x) = g(x)\, h(x)$ and if $g(x)$ and $h(x)$ are differentiable functions, then $f'(x) = g(x)\, h'(x) + h(x)\, g'(x)$. Hence it may be stated that the derivative of the product of two functions equals the first function times the derivative of the second plus the second function times the derivative of the first.

Proof Let $y = uv$, where u and v are functions of x. Then if x changes by an amount Δx, the corresponding changes in y, u, and v may be denoted by Δy, Δu, and Δv.

Hence $$y + \Delta y = (u + \Delta u)(v + \Delta v) = uv + u\,\Delta v + v\,\Delta u + \Delta u\,\Delta v$$
$$\Delta y = u\,\Delta v + v\,\Delta u + \Delta u\,\Delta v$$
$$\frac{\Delta y}{\Delta x} = u\frac{\Delta v}{\Delta x} + v\frac{\Delta u}{\Delta x} + \Delta u\frac{\Delta v}{\Delta x}$$
$$\lim_{\Delta x \to 0} \frac{\Delta y}{\Delta x} = u \lim_{\Delta x \to 0} \frac{\Delta v}{\Delta x} + v \lim_{\Delta x \to 0} \frac{\Delta u}{\Delta x} + \lim_{\Delta x \to 0} \Delta u \cdot \lim_{\Delta x \to 0} \frac{\Delta v}{\Delta x}$$

Hence $$\frac{dy}{dx} = u\frac{dv}{dx} + v\frac{du}{dx} + 0\frac{dv}{dx} = u\frac{dv}{dx} + v\frac{du}{dx}$$

Example 1:

$$f(x) = x^3(x-5)^{1/2}$$

$$f'(x) = x^3\left(\frac{1}{2}\right)(x-5)^{-1/2} + (x-5)^{1/2}(3x^2)$$

$$= \frac{x^3}{2(x-5)^{1/2}} + 3x^2(x-5)^{1/2}$$

$$= \frac{x^3 + 6x^2(x-5)}{2(x-5)^{1/2}} = \frac{7x^3 - 30x^2}{2(x-5)^{1/2}}$$

Example 2:

$$f(x) = \frac{x}{x+4} = x(x+4)^{-1}$$

$$f'(x) = x(-1)(x+4)^{-2} + (x+4)^{-1}(1)$$

$$= -\frac{x}{(x+4)^2} + \frac{1}{x+4}$$

$$= \frac{-x + (x+4)}{(x+4)^2} = \frac{4}{(x+4)^2}$$

In Example 2 we obtained the derivative of a function that is the quotient of two other functions by expressing the quotient as a product, using a negative exponent, and then applying the formula for the derivative of a product. However, an alternative procedure is available in the form of a formula for the derivative of a function that is the quotient of two differentiable functions:

Theorem II. If $f(x) = g(x)/h(x)$, $g(x)$ and $h(x)$ are differentiable functions, and $h(x) \neq 0$, then

$$f'(x) = \frac{h(x)g'(x) - g(x)h'(x)}{[h(x)]^2}$$

Hence, the derivative of the quotient of two functions equals the denominator times the derivative of the numerator minus the numerator times the derivative of the denominator, all divided by the square of the denominator.

Proof This theorem can be proved using the product formula and the chain rule.

$$f(x) = \frac{g(x)}{h(x)} = [g(x)][h(x)]^{-1}$$

$$= \{g(x)[-h(x)]^{-2}\}h'(x) + [h(x)]^{-1}g'(x)$$

$$= \frac{-g(x)h'(x)}{[h(x)]^2} + \frac{g'(x)}{h(x)}$$

$$= \frac{h(x)g'(x) - g(x)h'(x)}{[h(x)]^2}$$

Example 3:

$$f(x) = \frac{x}{x+4}$$

$$f'(x) = \frac{(x+4)(1) - x(1)}{(x+4)^2} = \frac{4}{(x+4)^2}$$

Example 4:

$$f(x) = \frac{x^2}{(2x-5)^{1/2}}$$

$$f'(x)\ 6= \frac{(2x-5)^{1/2}(2x) - x^2(\frac{1}{2})(2x-5)^{-1/2}(2)}{2x-5}$$

$$= \frac{2x(2x-5) - x^2}{(2x-5)^{3/2}}$$

$$= \frac{3x^2 - 10x}{(2x-5)^{3/2}}$$

5.8 Derivatives of Transcendental Functions

Thus far we have discussed only the derivatives of algebraic functions, i.e., functions that can be defined by polynomial equations. We now discuss the differentiation of nonalgebraic, or **transcendental functions,** such as exponential, logarithmic, trigonometric and hyperbolic functions.

The derivatives of the functions $\sin x$ and $\log_a x$ may be obtained using the definition of the derivative (Section 5.2). The derivatives of all the other transcendental functions that have been defined above may then be obtained from these derivatives.

To find the derivative of the function $\sin x$ we start with

$$\frac{d}{dx}(\sin x) = \lim_{\Delta x \to 0} \frac{\sin(x+\Delta x) - \sin x}{\Delta x}$$

However, the determination of this limit is not nearly as simple as it was in the case of algebraic functions. In obtaining the limit it is necessary to use another important limit:

$$\lim_{x\to 0} \frac{\sin x}{x} = 1$$

where x is assumed to be measured in radians. This is the reason that radian measure is used in the calculus of the trigonometric functions.

The important differentiation formula that results from the process described above is as follows:

$$\frac{d}{dx}(\sin x) = \cos x \tag{1}$$

By use of this formula it is relatively easy to establish the formulas for the other trigonometric functions. They are as follows:

$$\frac{d}{dx}(\cos x) = -\sin x \tag{2}$$

$$\frac{d}{dx}(\tan x) = \sec^2 x \tag{3}$$

$$\frac{d}{dx}(\csc x) = -\csc x \cot x \tag{4}$$

$$\frac{d}{dx}(\sec x) = \sec x \tan x \tag{5}$$

$$\frac{d}{dx}(\cot x) = -\csc^2 x \tag{6}$$

For the logarithmic function, carrying out the basic process for finding the derivative involves the use of another important limit:

$$\lim_{x\to 0} (1+x)^{1/x} = e$$

where e is the irrational number referred to in the discussion of the logarithmic function (Section 2.14). The differentiation formula obtained in this manner is as follows:

$$\frac{d}{dx}(\log_a x) = \frac{1}{x}\log_a e \tag{7}$$

If $a = e$, since $\log_e e = 1$ (recalling that $\log_e x$ is denoted by $\ln x$, we have

$$\frac{d}{dx}(\ln x) = \frac{1}{x} \tag{8}$$

The simplicity of (8) motivates the use of natural logarithms in the calculus.

Using the fact that the exponential function is the inverse of the logarithmic function, we arrive at the following differentiation formula:

$$\frac{d}{dx}(a^x) = a^x \ln a \tag{9}$$

In the important special case in which $a = e$, this formula becomes

$$\frac{d}{dx}(e^x) = e^x \tag{10}$$

Hence the function e^x has the remarkable property of being its own derivative.

The chain rule can be used to generalize these formulas. For example, if u is a differentiable function of x, Equations (1) and (8), respectively, become

$$\frac{d}{dx}(\sin u) = \cos u \frac{du}{dx}$$

$$\frac{d}{dx}(\ln u) = \frac{1}{u}\frac{du}{dx}$$

The other formulas, including others yet to be stated, can be generalized in the same way.

Before proceeding to some additional differentiation formulas for transcendental functions, we shall present some examples.

Example 1: $(u = 3x^2 - 5)$

$$\begin{aligned} f(x) &= \cos(3x^2 - 5) \\ f'(x) &= [-\sin(3x^2 - 5)](6x) \\ &= -6x \sin(3x^2 - 5) \end{aligned}$$

Example 2:

$$f(x) = \tan^2 4x$$
$$f'(x) = (2 \tan 4x)(\sec^2 4x)(4)$$
$$= 8 \tan 4x \sec^2 4x$$

Example 3:

$$f(x) = \csc x \cot x$$
$$f'(x) = (\csc x)(-\csc^2 x) + (\cot x)(-\csc x \cot x)$$
$$= -\csc^3 x - \csc x \cot^2 x$$
$$= -\csc x (\csc^2 x + \cot^2 x)$$

Example 4: $(u = 4x^2 - 1)$

$$f(x) = \ln(4x^2 - 1)$$
$$f'(x) = \left(\frac{1}{4x^2 - 1}\right)(8x)$$
$$= \frac{8x}{4x^2 - 1}$$

Example 5:

$$f(x) = \ln\left(\frac{3x + 5}{6x - 1}\right)$$

Solution The differentiation of this function can be simplified by first using one of the properties of the logarithmic function to obtain

$$f(x) = \ln(3x + 5) - \ln(6x - 1)$$

Hence

$$f'(x) = \frac{3}{3x + 5} - \frac{6}{6x - 1}$$
$$= \frac{3(6x - 1) - 6(3x + 5)}{(3x + 5)(6x - 1)}$$
$$= -\frac{33}{(3x + 5)(6x - 1)}$$

Example 6: $(u = 3x^2 - 5)$

$$f(x) = e^{3x^2 - 5}$$
$$f'(x) = (e^{3x^2 - 5})(6x)$$
$$= 6x e^{3x^2 - 5}$$

We shall now state the formulas for the derivatives of some of the other transcendental functions that have been discussed.

$$\frac{d}{dx}(\arcsin x) = \frac{1}{\sqrt{1-x^2}} \tag{11}$$

$$\frac{d}{dx}(\arccos x) = -\frac{1}{\sqrt{1-x^2}} \tag{12}$$

$$\frac{d}{dx}(\arctan x) = \frac{1}{1+x^2} \tag{13}$$

$$\frac{d}{dx}(\sinh x) = \cosh x \tag{14}$$

$$\frac{d}{dx}(\cosh x) = \sinh x \tag{15}$$

$$\frac{d}{dx}(\tanh x) = \text{sech}^2\, x \tag{16}$$

Example 7: $(u = x^3)$

$$f(x) = \arcsin(x^3)$$

$$f'(x) = \left(\frac{1}{\sqrt{1-x^6}}\right)(3x^2)$$

$$= \frac{3x^2}{\sqrt{1-x^6}}$$

Example 8: $(u = e^x)$

$$f(x) = \arctan e^x$$

$$f'(x) = \left(\frac{1}{1+(e^x)^2}\right)(e^x)$$

$$= \frac{e^x}{1+e^{2x}}$$

Example 9:

$$f(x) = x^2 \sinh x$$

$$f'(x) = x^2 \cosh x + 2x \sinh x$$

Example 10: $(u=\sqrt{x})$

$$f(x) = \cosh\sqrt{x}$$

$$f'(x) = (\sinh\sqrt{x})\frac{1}{2\sqrt{x}}$$

$$= \frac{1}{2\sqrt{x}}\sinh\sqrt{x}$$

5.9 Equations of Tangent Lines

It has already been shown that the derivative has an important geometric interpretation. If $f'(x)$ is the derivative of a function $f(x)$, then $f'(a)$ is the slope of the tangent to the graph of the function at the point $(a, f(a))$. We now use this fact, along with the point-slope form of the straight-line equation (Section 2.9) to obtain the equation of the tangent line to a curve at a given point.

Example 1 Find the equation of the tangent line to the graph of $f(x) = \sqrt{x}$ at the point at which $x = 4$.
Solution For our purposes the point-slope form of the straight-line equation may be written

$$[y - f(a)] = f'(a)(x - a)$$

In this case we have, since $a = 4$, the equations $f(4) = 2$, $f'(x) = 1/2\sqrt{x}$, $f'(4) = \frac{1}{4}$. Hence, the required equation is

$$(y-2) = \tfrac{1}{4}(x-4)$$

or

$$x - 4y + 4 = 0$$

Example 2 Find the equation of the tangent line to the graph of $f(x) = x \ln x$ at the point at which $x = e$.
Solution Here we have

$$f(e) = e \ln e = e$$

$$f'(x) = (x)\left(\frac{1}{x}\right) + \ln x = 1 + \ln x$$

$$f'(e) = 1 + 1 = 2$$

Hence the required equation is

$$(y - e) = 2(x - e)$$

or

$$2x - y - e = 0$$

5.10 Extremes of Functions

We now proceed to a second and a most important application of the derivative, i.e., the determination of the **extremes** of a function. The extremes of a function may be subdivided into **maximum** and **minimum** values of a function. These may further be classified as **relative** and **absolute** maximum or minimum values.

We confine our discussion initially to functions whose domain is an open interval or a union of open intervals. A function $f(x)$ is said to have a relative maximum at $x = a$ if $f(a) > f(x)$ for all x except $x = a$ in some open interval containing the value $x = a$. A function $f(x)$ is said to have a relative minimum at $x = a$ if $f(a) < f(x)$ for all x except $x = a$ in some open interval containing the value a.

A function is said to have an absolute maximum at $x = a$ if $f(a) > f(x)$ for all x except $x = a$ in the domain of the function. A function is said to have an absolute minimum at $x = a$ if $f(a) < f(x)$ for all x except $x = a$ in the domain of the function.

In Figure 5.3 there is shown the graph of a function $f(x)$ whose domain is the interval (a,b). The points A and C correspond to relative maxima of the function and are called maximum points. Point B corresponds to a relative minimum of the function and is called a minimum point. Evidently the point A also corresponds to an absolute maximum.

The graph in Figure 5.3 also suggests a geometric property of points such as A, B, and C, namely, that at such points, the tangent to the curve is a horizontal line, i.e., that the slope of the tangent

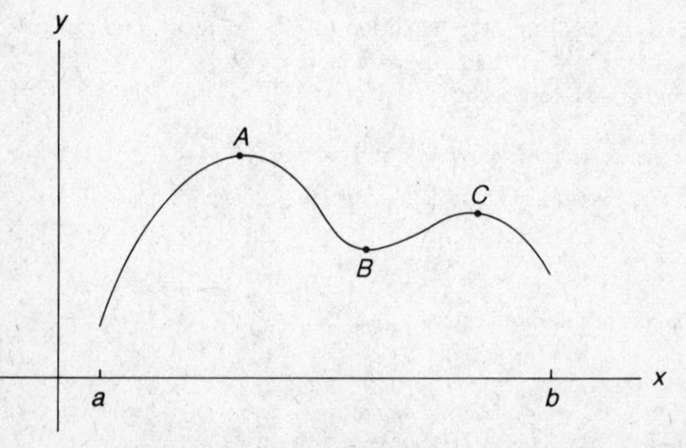

FIGURE 5.3

is 0. Points such as these can therefore be located by equating the first derivative of a function $f(x)$ to 0 and solving the resulting equation. The values of x thus obtained are called **critical values,** and the corresponding points on the curve are called **critical points.** The graph also indicates that at a maximum point such as A the slope of the tangent to the curve is positive just to the left of A and negative just to the right of A. More precisely, it may be stated that a critical point of the type under discussion, i.e., one at which the first derivative is equal to 0, is a maximum point if and only if in some open interval containing the critical value x_1, the first derivative is positive for all $x < x_1$ and negative for all $x > x_1$. A similar statement for minimum points holds if the word "positive" and "negative" are interchanged. We may summarize the above results by stating that:

(1) At a maximum point the derivative of the function changes from $+$ to $-$.

(2) At a minimum point the derivative of the function changes from $-$ to $+$.

Example 1 Find the extremes of the function

$$f(x) = \tfrac{1}{3}x^3 - \tfrac{3}{2}x^2 - 10x + 7$$

Solution Differentiating, we obtain

$$f'(x) = x^2 - 3x - 10$$

Hence to find the critical values we solve the equation

$$x^2 - 3x - 10 = 0$$
$$(x+2)(x-5) = 0$$
$$x = -2,\ x = 5$$

If we examine the derivative in the factored form, it is easy to see that

$$f'(x) \text{ is } \begin{cases} > 0 \text{ for } x < -2 \\ < 0 \text{ for } -2 < x < 5 \\ > 0 \text{ for } x > 5 \end{cases}$$

Hence from our discussion in the preceding paragraph, it is clear that $f(-2) = 55/3$ is a relative maximum and $f(5) = -233/6$ is a relative minimum. It is evident from the graph in Figure 5.4 that this function has no absolute extremes.

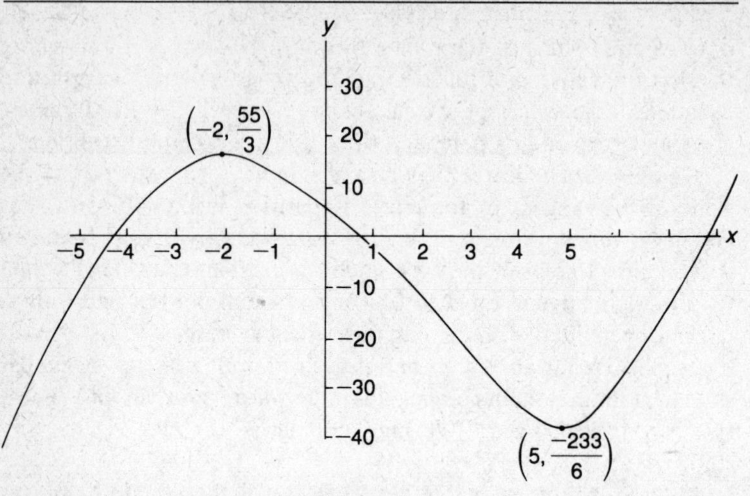

FIGURE 5.4

Example 2:

$$f(x) = xe^{-x}$$

Solution Differentiating, we obtain

$$\begin{aligned} f'(x) &= (x)(-e^{-x}) + e^{-x} \\ &= e^{-x}(1 - x) \end{aligned}$$

Let $e^{-x}(1 - x) = 0$. Since $e^{-x} \neq 0$ for all x, the only solution is

$$x = 1$$

Now

$$f'(x) \text{ is } \begin{cases} > 0 \text{ for } x < 1 \\ < 0 \text{ for } x > 1 \end{cases}$$

Hence, $f(1) = e^{-1}$ is a relative maximum. It is also an absolute maximum since the above analysis of the signs of $f'(x)$ shows that $f(x)$ is increasing for all $x < 1$ and decreasing for all $x > 1$. The graph of the function is shown in Figure 5.5.

The fact that $f'(x)$ is 0 for a certain value of x does not ensure the occurrence of a maximum or minimum value for $f(x)$, as is shown in the following example.

Example 3:

$$\begin{aligned} f(x) &= x^3 \\ f'(x) &= 3x^2 \end{aligned}$$

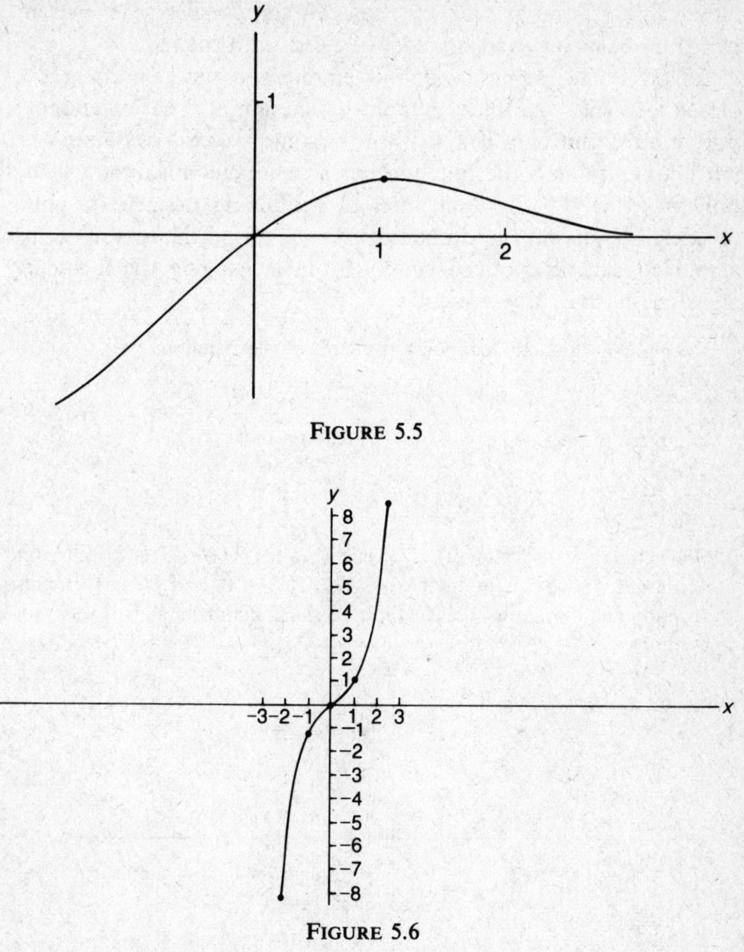

FIGURE 5.5

FIGURE 5.6

Let $$3x^2 = 0$$

Then $$x = 0$$

Hence $x = 0$ is a critical value, but $f'(x) > 0$ for all $x \neq 0$, so $f(0)$ is neither a maximum nor a minimum. The graph of $f(x)$ is shown in Figure 5.6.

Thus far our discussion of extremes has been confined to functions whose domains are open intervals or unions of open intervals. We shall now consider functions whose domains include one or more closed intervals. For such functions an additional type of extreme called an **endpoint extreme** must be taken into account.

As the name implies, endpoint extremes occur at endpoints of closed intervals. At a left endpoint a function $f(x)$ has an endpoint minimum if and only if $f'(x) > 0$ in some interval containing the left endpoint,* and the function has an endpoint maximum if and only if $f'(x) < 0$ in some interval containing the left endpoint. At a right endpoint the situation is reversed, and the corresponding statement can be obtained simply by interchanging the inequality signs in the preceding sentence.

Example 4 Find the endpoint extremes of the function

$$f(x) = 4x - x^2 \qquad x \in [1,4]$$

Solution Here $f'(x) = 4 - 2x$. Furthermore

$$f'(x) \text{ is } \begin{cases} > 0 & x < 2 \\ < 0 & x > 2 \end{cases}$$

Hence in the interval (1,2), $f'(x) > 0$, and $f(1) = 3$ is an endpoint minimum. Also, in the interval (2,4), $f'(x) < 0$, and $f(4) = 0$ is also an endpoint minimum. The graph of the function is shown in Figure 5.7.

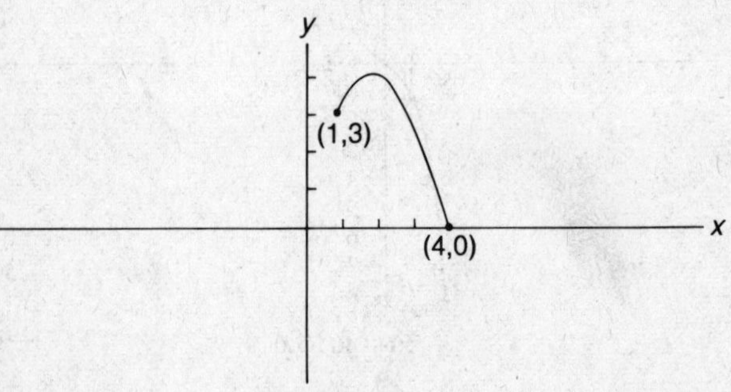

FIGURE 5.7

There remains only one additional type of extreme, which we shall mention in passing. A function $f(x)$ may have an extreme value for a value of x at which $f'(x)$ does not exist, or in geometrical terms, at a point at which the graph of the function has a vertical tangent. However, this case has little or no practical application.

In determining whether a function $f(x)$ has a maximum or a minimum value for a value x_1 at which $f'(x_1) = 0$, we have used what

* Except at the endpoint itself where $f'(x)$ does exist.

may be called the first-derivative test; i.e., we have examined the sign of $f'(x)$ for values of x close to x_1. However, a second procedure, which may be called the second-derivative test, may be applied to a function $f(x)$, if $f'(x_1) = 0$. If $f''(x_1) < 0$, then $f(x_1)$ is a maximum; whereas if $f''(x_1) > 0$, then $f(x_1)$ is a minimum. If $f''(x_1) = 0$, then the test is inconclusive.

Example 5 Determine the extremes of the function $f(x) = x^3 - 3x^2$, using the second-derivative test.

$$f'(x) = 3x^2 - 6x$$

$$f''(x) = 6x - 6$$

Let $$3x^2 - 6x = 0$$

$$3x(x-2) = 0$$

Hence $$x = 0, \ x = 2$$

Then $$f''(0) = -6 < 0$$

Therefore $$f(0) = 0 \quad \text{is a maximum}$$

Also $$f''(2) = 6 > 0$$

Therefore $$f(2) = -4 \quad \text{is a minimum}$$

We conclude this discussion of maxima and minima with two practical applications.

Example 6 A farmer wishes to enclose a rectangular area adjacent to a barn by fencing in three sides of the enclosure. If 120 ft of fencing are to be used, what dimensions will produce the maximum area?

Solution Designate by x the length of the side of the enclosure parallel to the barn and by y the length of each of the sides perpendicular to the barn. Then the area of the enclosure is the product xy. Since x and y are related by the equation $x + 2y = 120$, we can write $y = \frac{1}{2}(120 - x)$. Hence, if we denote the area of the enclosure by $a(x)$, we have

$$a(x) = \tfrac{1}{2}x(120 - x)$$

$$= 60x - \tfrac{1}{2}x^2$$

$$a'(x) = 60 - x$$

Let $$60 - x = 0$$

then $$x = 60$$

Furthermore, since $a''(x) = -1$, $a''(x) < 0$ for all x. Hence $a(60) = 1800$ is the maximum area for the enclosure, and since for $x = 60$, $y = \frac{1}{2}(120 - 60) = 30$, the dimensions for maximum area are 60 ft $\times$ 30 ft.

Example 7 A rectangular box with an open top is to be made with a base whose length equals twice its width. If the volume of the box is to be 2160 cm^3, what dimensions will minimize the surface area (and therefore the amount of material to be used)?

Solution Designate the length of the base by $2x$, the width by x, and the depth of the box by y. Then the surface area, which is the sum of the area of the base and the areas of the four sides, is $2x^2 + 6xy$. Since the volume of the box is to be 2160 cm^3, x and y are related by the equation $2x^2y = 2160$ or $y = 1080/x^2$. Hence, if the surface area is designated by $s(x)$, we have

$$s(x) = 2x^2 + 6x\left(\frac{1080}{x^2}\right)$$

$$= 2x^2 + \frac{6480}{x}$$

Hence $$s'(x) = 4x - \frac{6480}{x^2}$$

Let $$4x - \frac{6480}{x^2} = 0$$

or $$4x^3 - 6480 = 0$$

$$x^3 = 1620$$

$$x = \sqrt[3]{1620} \simeq 11.74 \text{ cm}$$

Furthermore $$s''(x) = 4 + \frac{12{,}960}{x^3}$$

Evidently $$s'(\sqrt[3]{1620}) > 0$$

Hence, $s(\sqrt[3]{1620})$ is the minimum volume for the box. Using the relationship $y = 1080/x^2$, we obtain as the approximate dimensions of the box 11.74 cm $\times$ 23.48 cm $\times$ 7.84 cm.

5.11 Velocity and Acceleration

Consider a particle that moves along a horizontal coordinate line. Its position x at any time t is given by a specified function $x(t)$. Then the first derivative $x'(t)$ is the rate of change of position, or the velocity of the particle, and the second derivative $x''(t)$ is the rate of change of velocity, or the acceleration of the particle.

Example 1 A particle moves along a coordinate line so that its position is given by $x(t) = t^2 - 4t$, $t \geq 0$, where x is measured in centimeters and t in seconds.

(a) What is the initial position of the particle?

(b) What is its initial velocity?

(c) During what time interval is the particle moving to the left?

(d) What is the point farthest to the left the particle will reach?

(e) What is the acceleration of the particle at time t?

Solution (a) The initial position of the particle is $x(0) = 0$, i.e., the particle starts at the origin of the coordinate line.
Solution (b) The velocity of the particle at time t is $x'(t) = 2t - 4$, and the initial velocity is $x'(0) = -4$ cm/sec. The positive direction being to the right, the negative velocity indicates that the particle is moving to the left.
Solution (c) Since $2t - 4 < 0$ for $0 < t < 2$, the particle moves to the left during that time interval.
Solution (d) Since the particle will move to the right for all $t > 2$, the point farthest left that will be reached is $x(2) = -4$, i.e., a point 4 cm to the left of the origin.
Solution (e) The acceleration at any time is $x''(t) = 2$; the acceleration of the particle is constant.

It may be shown that if an object is released at height y_0 above the ground with initial velocity v_0 and allowed to fall freely under the force of gravity, air resistance being neglected, the height y of the object in feet after t sec is given by

$$y = -16t^2 + v_0 t + y_0$$

In determining this equation, the upward direction is taken as positive and the gravitational constant is taken as 32.

Example 2 An object is released from a height of 256 ft with an initial upward velocity of 96 ft/sec.

(a) When will the object strike the ground?

(b) With what velocity will it strike the ground?

(c) What is the greatest height reached by the object?

Solution (a) In this case, we have $v_0 = 96$ and $y_0 = 256$, and the height of the object at time t will be

$$y = -16t^2 + 96t + 256$$

Hence we set

$$-16t^2 + 96t + 256 = 0$$

or
$$t^2 - 6t - 16 = 0$$
$$(t-8)(t+2) = 0$$
$$t = 8,\ t = -2$$

The root -2 is evidently extraneous to the physical problem, and we conclude that the object will strike the ground after 8 sec.

Solution (b) We first obtain the velocity at time t by calculating $y' = -32t + 96$. Then the velocity with which the object will strike the ground is

$$y'(8) = -160 \text{ ft/sec}$$

where the negative sign indicates that the direction is downward.

Solution (c) We note that the greatest height will be reached at the instant the velocity is zero. Hence we set

$$-32t + 96 = 0$$

Then
$$t = 3$$

Hence the greatest height will be reached after 3 sec, and the maximum height will be $y(3) = 400$ ft.

5.12 Rate of Change

We also use the derivative and, in particular, the chain rule, to determine rates of change involving more than two variables. If y is a function of x and x is a function of t, then, by the chain rule, we have

$$\frac{dy}{dt} = \frac{dy}{dx}\frac{dx}{dt}$$

Example 1 A particle moves along the curve $y = x^3 - 2x$ in such a way that its abscissa (x coordinate) is increasing at the constant rate of 4 units/sec. What is the rate of change of the ordinate (y coordinate) at the instant when the particle passes through the point (2,4)? Here

$$\frac{dy}{dx} = 3x^2 - 2 \qquad \text{and} \qquad \frac{dx}{dt} = 4$$

Using the chain rule yields

$$\frac{dy}{dt} = (3x^2 - 2)(4) = 12x^2 - 8$$

Hence, at the instant when the particle passes through the point (2,4), we have

$$\left.\frac{dy}{dt}\right|_{x=2} = 40 \text{ units/sec}$$

Example 2 A spherical ballon is inflated in such a way that the radius is increasing at the rate of 5 cm/sec. How fast is the volume increasing at the instant when the radius is 10 cm long?

Solution The volume of a sphere of radius r is given by

$$v = \frac{4}{3}\pi r^3$$

Hence
$$\frac{dv}{dr} = 4\pi r^2 \qquad \frac{dr}{dt} = 5$$

Hence
$$\frac{dv}{dt} = \frac{dv}{dr}\frac{dr}{dt}$$

$$= (4\pi r^2)(5) = 20\pi r^2$$

and when
$$r = 10,\ dv/dt = 2000\pi \text{ cm}^3/\text{sec}$$

5.13 Use of the Derivative in the Determination of Limits

In Section 2.2, we discussed the limit of a function that was of the form $g(x)/h(x)$ and for which $g(x)$ and $h(x)$ both increase without limit as $x \to \infty$. The limit was obtained by an algebraic modification of the function. However, the limits, if they exist, of many functions of this type cannot be obtained in this manner. In such cases the derivative provides a useful means of determining the limit.

It can be shown that under certain broad conditions, which will hold for all of the functions considered in this book, the limit, if it exists, of a function of the type $f(x)/g(x)$, for which both $f(x)$ and $g(x)$ increase without limit, is the same as the limit, if it exists, of the corresponding quotient of the derivatives of $f(x)$ and $g(x)$. More specifically, if as $x \to a$ or $x \to \infty$, $f(x) \to \infty$ and $g(x) \to \infty$, then

$$\lim \frac{f(x)}{g(x)} = \lim \frac{f'(x)}{g'(x)}$$

In Section 2.2 we found

$$\lim_{x\to\infty} \frac{4x-3}{2x+5} = 2$$

by algebraic means. Note that the same result is obtained by using derivatives:

$$\lim_{x\to\infty}\frac{4x-3}{2x+5}=\lim_{x\to\infty}\frac{\frac{d}{dx}(4x-3)}{\frac{d}{dx}(2x+5)}$$

$$=\lim_{x\to\infty}\frac{4}{2}=2$$

Example 1 Find $\lim_{x\to\infty}\frac{x}{e^x}$

$$\lim_{x\to\infty}\frac{x}{e^x}=\lim_{x\to\infty}\frac{\frac{d}{dx}(x)}{\frac{d}{dx}(e^x)}=\lim_{x\to\infty}\frac{1}{e^x}=0$$

Example 2:

$$\lim_{x\to\infty}\frac{\ln x}{x}=\lim_{x\to\infty}\frac{\frac{1}{x}}{1}=\lim_{x\to\infty}\frac{1}{x}=0$$

The same procedure may be employed in cases in which both $f(x)$ and $g(x)$ approach 0 as a limit. Consider the following.

Example 3 Find $\lim_{x\to\pi/2}\frac{1-\sin x}{\cos x}$

$$\lim_{x\to\pi/2}\frac{1-\sin x}{\cos x}=\lim_{x\to\pi/2}\frac{\frac{d}{dx}(1-\sin x)}{\frac{d}{dx}(\cos x)}$$

$$=\lim_{x\to\pi/2}\frac{-\cos x}{-\sin x}=\lim_{x\to\pi/2}\frac{0}{-1}=0$$

5.14 Integration

In this chapter thus far we have discussed various aspects of the problem, given a function $f(x)$, find its derivative $f'(x)$. Here we discuss the reverse problem, given a function $f(x)$, find a function $F(x)$ such that $F'(x)=f(x)$; i.e., find a function whose derivative is the given function.

Suppose, for example, that the given function is $f(x) = x^2$. What is the function $F(x)$ such that $F'(x) = x^2$? From our knowledge of differentiation it appears that the desired function should be a multiple of x^3. Furthermore, since $d/dx(x^3) = 3x^2$, it is easily seen that the function $F(x) = \frac{1}{3}x^3$ is a solution to our problem since $F'(x) = \frac{1}{3}(3x^2) = x^2$.

The function $\frac{1}{3}x^3$ is said to be an **antiderivative** of the function x^2. However it is readily seen that the function $\frac{1}{3}x^3$ is by no means the only antiderivative of the function x^2. For example, the functions $g(x) = \frac{1}{3}x^3 + 5$ and $h(x) = \frac{1}{3}x^3 - 100$ are also functions whose derivatives are equal to x^2. In fact, any function of the form $\frac{1}{3}x^3 + C$, where C is any constant, will have x^2 for its derivative.

The operation by which the function $\frac{1}{3}x^3 + C$ was obtained from the function x^2 in the above example is known as integration, and the relationship between the two functions is denoted symbolically as follows:

$$\int x^2\,dx = \frac{1}{3}x^3 + C$$

The symbol $\int dx$ therefore denotes the operation of integration just as the symbol d/dx denotes the operaton of differentiation. The symbol $\int$ is referred to as the integral sign.

Hence, if $F(x)$ is an antiderivative of $f(x)$, then

$$\int f(x)\,dx = F(x) + C$$

The function $F(x) + C$ is called the **indefinite integral** of the function $f(x)$. The constant C, which is called an arbitrary constant since it can assume any value, is referred to as the **constant of integration.**

It can be shown that the indefinite integral, as defined above, is completely general; i.e., if $F(x)$ is any antiderivative of $f(x)$, then all antiderivatives of $f(x)$ are of the form $F(x) + C$.

Two important properties of integrals that follow immediately from the corresponding properties of derivatives are listed below.

$$\int kf(x)\,dx = k\int f(x)\,dx \qquad \text{where } k \text{ is a constant} \qquad (1)$$

$$\int [f(x) + g(x)]\,dx = \int f(x)\,dx + \int g(x)\,dx \qquad (2)$$

In the example discussed above, we found that $\frac{1}{3}x^3$ is an antiderivative of x^2. This leads to the conjecture that $x^{n+1}/(n+1)$ is an antiderivative of x^n. This is easily verified since

$$\frac{d}{dx}\left[\frac{x^{n+1}}{n+1}\right]=\frac{1}{n+1}(n+1)x^n=x^n$$

Note, however, that the symbol $x^{n+1}/(n+1)$ has no meaning when $n=-1$. Hence we have the integration formula

$$\int x^n\,dx=\frac{x^{n+1}}{n+1} \qquad n\neq -1$$

In particular, when $n=0$ we have

$$\int 1\,dx=\int x^0\,dx=\frac{x^1}{1}+C$$

Hence we have the formula

$$\int dx=x+C$$

These formulas as well as properties (1) and (2) above are illustrated in the following examples.

$$\int(4x^3+2x^2-5x+4)\,dx$$

$$=x^4+\frac{2x^3}{3}-\frac{5x^2}{2}+4x+C$$

$$\int\sqrt{3}x=\sqrt{3}\int x^{1/2}\,dx=\sqrt{3}\,\frac{x^{3/2}}{\frac{3}{2}}=\frac{2\sqrt{3}}{3}\,x^{3/2}+C$$

$$\int\frac{4}{x^2}=4\int x^{-2}\,dx=\frac{4x^{-1}}{-1}=-\frac{4}{x}+C$$

$$\int\frac{dx}{\sqrt[3]{x}}=\int x^{-1/3}\,dx=\frac{x^{2/3}}{\frac{2}{3}}+C=\tfrac{3}{2}\,x^{2/3}+C$$

$$\int(x-4)^3\,dx=\int(x^3-12x^2+48x-64)\,dx$$

$$=\frac{x^4}{4}-4x^3+24x^2-64x+C$$

5.15 Particular Integrals

An integral of the form $F(x) + C$ is indefinite in the sense that it contains an undetermined constant. However, in many problems involving integration, sufficient information is given to enable us to determine a specific value for C. In this case we obtain what is called a **particular integral.** Two illustrations of this type of problem will now be presented.

Example 1 Consider the problem of finding the equation of the curve $y = f(x)$ that satisfies the following conditions.

(a) The slope of the tangent to the curve at any point equals twice the x coordinate of the point.

(b) The curve passes through the point $(2,-1)$.

Solution In the discussion of the geometric interpretation of the derivative it was shown that the value of the derivative of a function at a particular point of the graph of the function is equal to the slope of the tangent to the curve at that point. Therefore in view of condition (a) above we have

$$\frac{dy}{dx} = f'(x) = 2x$$

Hence
$$y = \int f'(x)\,dx = \int 2x\,dx = x^2 + C$$

Condition (b) tells us that $y = -1$ when $x = 2$. Substituting these values in the above equation, we have $-1 = 4 + C$ or $C = -5$. The equation of the curve satisfying the given conditions is $y = x^2 - 5$.

Example 2 Find $f(x)$, given $f''(x) = 4x - 3$, $f'(0) = 4$, $f(0) = 3$.
Solution Since $f''(x)$ denotes the second derivative of the function $f(x)$, it is clear that the first derivative can be found by integration.

Hence $\quad f'(x) = \int f''(x)\,dx = \int (4x - 3)\,dx = 2x^2 - 3x + C_1$

Since $\quad f'(0) = 4$, we have $C_1 = 4$.

Hence $\quad f'(x) = 2x^2 - 3x + 4$

Therefore $\quad f(x) = \int f'(x)\,dx = \int (2x^2 - 3x + 4)\,dx$

$$= \frac{2}{3}x^3 - \frac{3}{2}x^2 + 4x + C_2$$

Since $\quad f(0) = 3$, we have $C_2 = 3$.

Therefore $$f(x) = \frac{2}{3}x^3 - \frac{3}{2}x^2 + 4x + 3$$

5.16 Integrals As Areas; The Definite Integral

We are already familiar with procedures for finding the areas of plane figures such as rectangles, triangles, and trapezoids. We consider at this point the more complicated problem of calculating an area such as the one shown in Figure 5.8, an area bounded by the x axis, the vertical lines $x = a$ and $x = b$, and the graph of the function, $y = f(x)$.

(More precisely we might refer to the area of the *region* bounded by the x axis, etc. However, the briefer language will be used in the discussion of areas since the meaning is clear.)

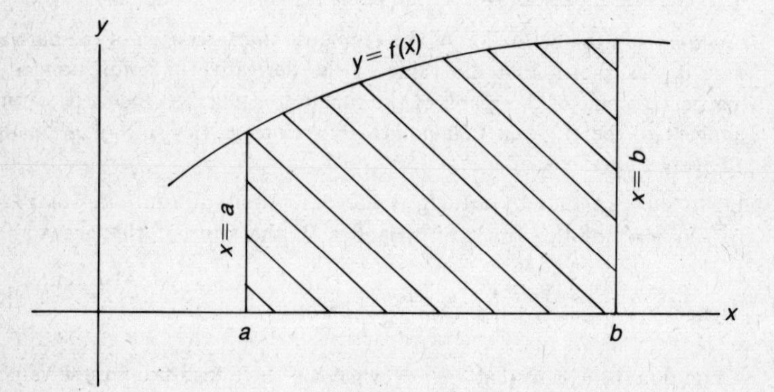

FIGURE 5.8

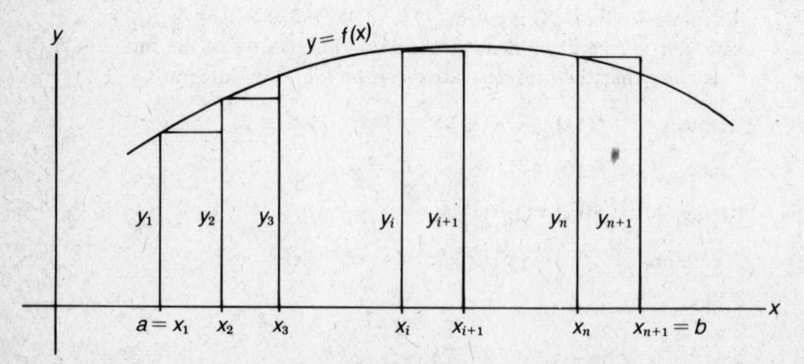

FIGURE 5.9

Although none of the well-known geometrical formulas for calculating plane areas will suffice to determine the area exactly, the area can be approximated by constructing a number of rectangles and using their total area to approximate the desired area, as illustrated in Figure 5.9.

The interval $[a,b]$ is divided into n subintervals by the insertion of $n - 1$ points of subdivision, $x_2, x_3, \ldots, x_i, x_{i+1}, \ldots, x_n$. For convenience of notation let $a = x_1$ and $b = x_{n+1}$. The length of the ith subinterval is $\Delta x_i = x_{i+1} - x_i$. In each subinterval a rectangle having the subinterval as base and the ordinate to the curve at the left endpoint of the subinterval as altitude is drawn. The area of the ith rectangle is therefore $y_i \Delta x_i$.

Using summation notation, the sum of the areas of the n rectangles can be written

$$\sum_{i=1}^{n} y_i \Delta x_i = \sum_{i=1}^{n} f(x_i) \Delta x_i$$

Evidently, the sum approximates the required area. It is also clear that the larger the number of subdivisions, the better the approximation will be. In fact by letting n increase without limit in such a way that each of the Δx_i's approaches 0, the sum of the areas of the rectangles can be brought as close as desired to the area in question.

As a matter of fact, the limit referred to above provides the basis for a precise definition of the area with which we are concerned. Denoting this area by A, we have

$$A = \lim_{\substack{n \to \infty \\ \Delta x_i \to 0}} \sum_{i=1}^{n} f(x_i) \Delta x_i$$

The relationship between the area defined above and the integral of the function $f(x)$ is given by an extremely important theorem known as the fundamental theorem of the integral calculus. It may be stated as follows.

Fundamental theorem. If $f(x)$ is a continuous function defined in the interval $[a,b]$, where the interval has been subdivided as shown above, and if $F(x)$ is an antiderivative of $f(x)$, then

$$\lim_{\substack{n \to \infty \\ \Delta x_i \to 0}} \sum_{i=1}^{n} f(x_i) \Delta x_i = F(b) - F(a)$$

The quantity on the left-hand side of the above equation is commonly denoted by the symbol

$$\int_a^b f(x)\,dx$$

This limit, which exists for all continuous functions, is called the **definite integral** of $f(x)$ from a to b. Hence

$$\int_a^b f(x)\,dx = F(b) - F(a)$$

This important formula shows that the definite integral may be evaluated by using an antiderivative of $f(x)$. The number a is called the lower limit, and the number b, the upper limit of the definite integral. The variable x in $\int_a^b f(x)\,dx$ is sometimes referred to as a **dummy variable** since it can be replaced with any other letter without changing the value of the integral, e.g.,

$$\int_a^b f(x)\,dx = \int_a^b f(t)\,dt = F(b) - F(a)$$

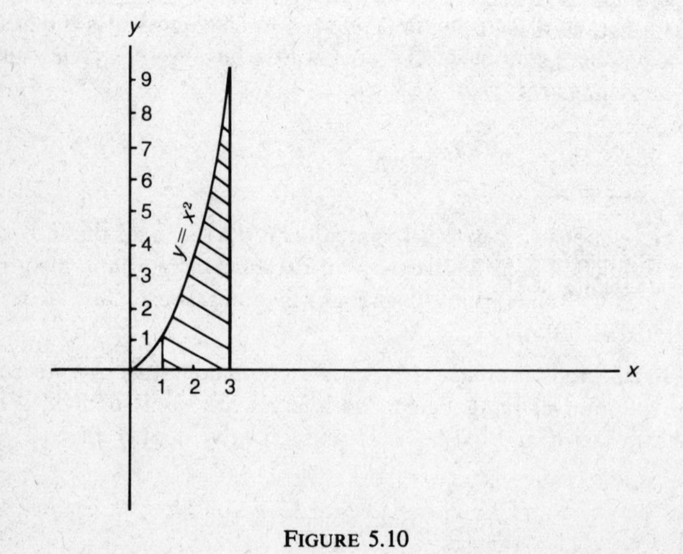

FIGURE 5.10

Hence we conclude that the area under consideration can be calculated by means of the definite integral, i.e.,

$$A = \int_a^b f(x)\,dx = F(b) - F(a)$$

The quantity $F(b) - F(a)$ is often indicated by $[F(x)]_a^b$.

Example 1 Consider the problem of finding the area bounded by the curve $y = x^2$, the x axis, and the lines $x = 1$ and $x = 3$. The required area is shown in Figure 5.10.

$$A = \int_1^3 x^2\,dx = \left[\frac{x^3}{3}\right]_1^3 = 9 - \frac{1}{3} = 8\tfrac{2}{3}$$

Example 2 Suppose it is desired to find the area bounded by the curve $y = \sqrt{x}$, the x axis, and the line $x = 4$. The area is shown in Figure 5.11.

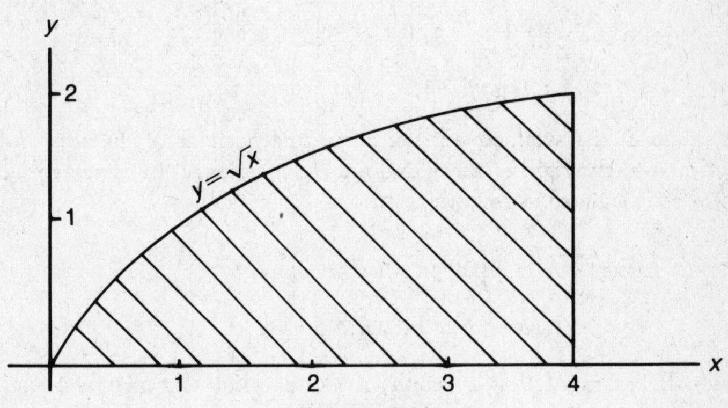

FIGURE 5.11

$$A = \int_0^4 x^{1/2}\,dx$$

$$= \left[\frac{2}{3}x^{3/2}\right]_0^4 = \frac{2}{3}[4^{3/2} - 0]$$

$$= \frac{2}{3} \cdot 8 = \frac{16}{3} = 5\tfrac{1}{3}$$

Example 3 Finally, consider the problem of finding the area bounded by the curve $y = 9 - x^2$ and the x axis. A sketch of the required area is shown in Figure 5.12.

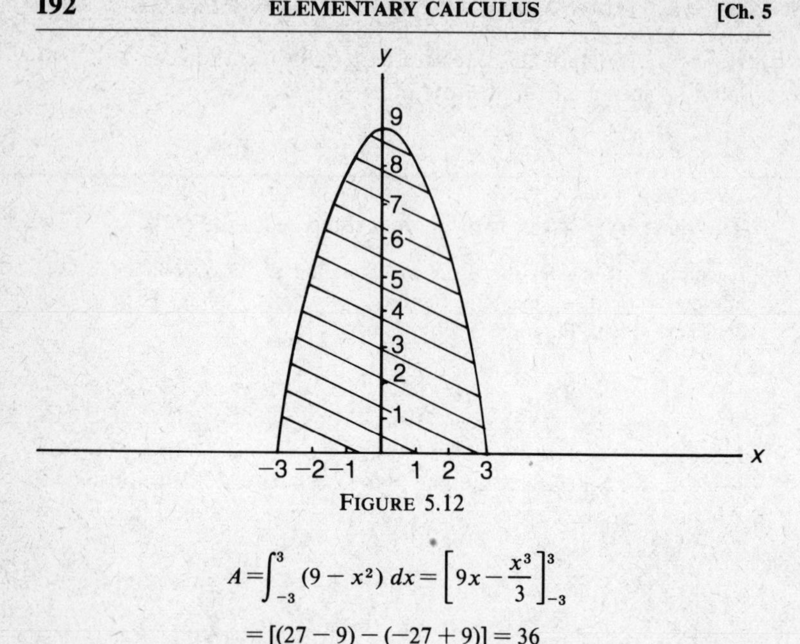

FIGURE 5.12

$$A = \int_{-3}^{3} (9 - x^2)\, dx = \left[9x - \frac{x^3}{3} \right]_{-3}^{3}$$

$$= [(27 - 9) - (-27 + 9)] = 36$$

Because of the symmetry of the curve, the portions of the area to the left and to the right of the y axis are equal. Hence the above area can also be calculated as follows:

$$A = 2\int_{0}^{3} (9 - x^2)\, dx = 2\left[9x - \frac{x^3}{3} \right]_{0}^{3}$$

$$= 2[(27 - 9) - (0)] = 36$$

It will be noted that in each of the examples shown above, the area to be calculated lies entirely above the x axis. It follows from the definition of the definite integral that if the procedure used in these examples were applied to the calculation of an area lying entirely below the x axis, a negative number would be obtained. Hence in cases in which the area to be calculated lies at least partially below the x axis, our procedure must be modified.

Example 4 Consider the problem of calculating the shaded area shown in Figure 5.13.

Here
$$A = \int_{a}^{b} f(x)\, dx + \left| \int_{b}^{c} f(x)\, dx \right|$$

$$= \int_{a}^{b} f(x)\, dx - \int_{b}^{c} f(x)\, dx$$

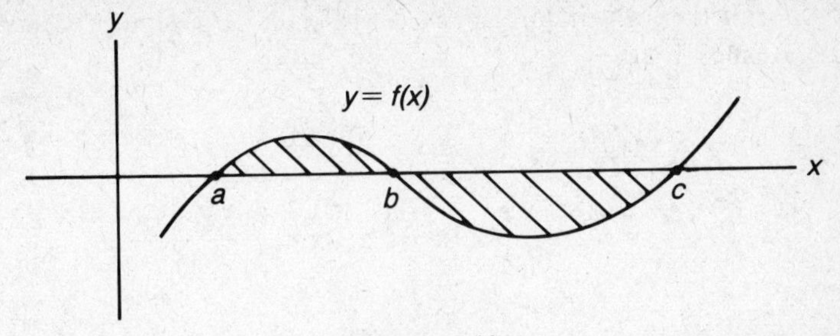

FIGURE 5.13

The calculation of areas as illustrated in the previous examples is by no means the only application of the definite integral. Using the fundamental theorem, the integral can be applied to the calculation of any quantity that can be expressed as the limit of a sum of the form $\Sigma f(x_i)\Delta x_i$. In this way the integral can be applied to many important geometric and physical problems, including the calculation of volumes, centroids, moments of inertia, work, fluid pressure, etc.

5.17 Further Procedures for Integrating Algebraic Functions

In one of the previous examples the function $f(x) = (x - 4)^3$ was integrated by first expanding the binominal and then integrating the resulting terms individually, using the formula for $\int x^n\, dx$. Clearly the same procedure could be employed to find $\int (3x - 5)^{10}\, dx$, but the calculation would be quite tedious. Furthermore, the calculation of an integral such as $\int \sqrt{2x-5}\, dx$ could not be carried out by this procedure since the number of terms in the binomial expansion of $(2x-5)^{1/2}$ is infinite. In order to handle the integration of functions of this kind further techniques are required, and these will now be discussed.

To return to the problem of obtaining $\int (3x - 5)^{10}\, dx$, in view of the rule for integrating x^n we might conjecture that $(3x - 5)^{11}/11$ would be an antiderivative of $(3x - 5)^{10}$. This conjecture is easily shown to be incorrect since if $y = (3x - 5)^{11}/11$, we find, using the chain rule, that

$$\frac{dy}{dx} = \frac{1}{11} \cdot 11(3x-5)^{10} \cdot 3 = 3(3x-5)^{10}$$

We have therefore shown that $(3x - 5)^{11}/11$ is an antiderivative of the function $3(3x-5)^{10}$. Since this function is a constant multiple of the function whose integral is desired, the following procedure can be employed:

$$\int (3x-5)^{10}\,dx = \int \frac{1}{3}\cdot 3(3x-5)^{10}\,dx$$

$$= \frac{1}{3}\int 3(3x-5)^{10}\,dx = \frac{1}{3}\frac{(3x-5)^{11}}{11} + C$$

$$= \frac{(3x-5)^{11}}{33} + C$$

The procedure used in the above example can be formalized as follows. Let $u = 3x - 5$, and denote by du the quantity

$$\frac{d}{dx}(3x-5)\,dx$$

Hence $du = 3\,dx$. Therefore

$$\int (3x-5)^{10}\,dx = \frac{1}{3}\int (3x-5)^{10}(3\,dx)$$

$$= \frac{1}{3}\int u^{10}\,du = \frac{1}{3}\frac{u^{11}}{11} + C = \frac{(3x-5)^{11}}{33} + C$$

The same procedure can be used to calculate $\int \sqrt{2x-5}\,dx$. In this case let $u = 2x - 5$. Then

$$du = \frac{d}{dx}(2x-5)\,dx = 2\,dx$$

Therefore

$$\int \sqrt{2x-5}\,dx = \frac{1}{2}\int (2x-5)^{1/2}(2\,dx)$$

$$= \frac{1}{2}\int u^{1/2}\,du = \frac{1}{2}\frac{u^{3/2}}{\frac{3}{2}} + C$$

$$= \frac{1}{3}u^{3/2} + C = \frac{1}{3}(2x-5)^{3/2} + C$$

For a third example consider $\int x(x^2 - 5)^6\,dx$. In this case let $u = x^2 - 5$. Then $du = 2x\,dx$. Hence

$$\int x(x^2-5)^6\,dx=\frac{1}{2}\int (x^2-5)^6(2x\,dx)$$

$$=\frac{1}{2}\int u^6\,du=\frac{1}{2}\frac{u^7}{7}=\frac{(x^2-5)^7}{14}+C$$

Note that in each of the above examples a needed constant was supplied after the integral sign and the reciprocal of that constant was placed before the integral sign.

The procedure employed in these examples can be generalized as follows.

Suppose $f(x)$ is a function that can be expressed as $f(x) = [u(x)]^n \cdot g(x)$, where n is any constant except -1. Calculate $du = u'(x)\,dx$. If possible, find a constant k such that $kg(x)\,dx = du$. Then

$$\int f(x)\,dx=\int [u(x)]^n g(x)\,dx$$

$$=\frac{1}{k}\int [u(x)]^n \cdot kg(x)\,dx$$

$$=\frac{1}{k}\int u^n\,du=\frac{1}{k}\frac{u^{n+1}}{n+1}+C$$

This general procedure will now be illustrated in the calculation of

$$\int \frac{x^2\,dx}{(x^3-4)^3}$$

Let $u(x) = x^3 - 4$, $g(x) = x^2$, $n = -3$. Then $u'(x) = 3x^2$ and $du = 3x^2\,dx$. Hence $k = 3$ since $3g(x)\,dx = du$. Therefore

$$\int \frac{x^2\,dx}{(x^3-4)^3}=\frac{1}{3}\int (x^3-4)^{-3}(3x^2\,dx)$$

$$=\frac{1}{3}\int u^{-3}\,du=\frac{1}{3}\frac{u^{-2}}{-2}+C$$

$$=\frac{1}{3}\frac{(x^3-4)^{-2}}{-2}+C=-\frac{1}{6(x^3-4)^2}+C$$

Note that if $u=f(x)$ and $du=f'(x)dx$, the formula

$$\int u^n\,du = \frac{u^{n+1}}{n+1} + C \qquad n \neq -1$$

may also be written in the form

$$\int [f(x)]^n f'(x)\,dx = \frac{[f(x)]^{n+1}}{n+1} + C \qquad n \neq -1$$

5.18 Integrals Involving Transcendental Functions

It will be recalled that for the exponential function $f(x) = e^x$ we have $f'(x) = e^x$. Hence e^x is an antiderivative of itself and we have

$$\int e^x\,dx = e^x + C$$

It is readily seen however that the function e^{2x} does not have this property since

$$\frac{d}{dx}\,e^{2x} = 2e^{2x}$$

Hence e^{2x} is an antiderivative of $2e^{2x}$. Hence

$$\int e^{2x}\,dx = \int \frac{1}{2}\cdot 2e^{2x}\,dx$$

$$= \frac{1}{2}\int 2e^{2x}\,dx = \frac{1}{2}\,e^{2x} + C$$

The procedure used above is evidently similar to the method used previously to integrate certain algebraic functions; i.e., a constant was introduced after the integral sign, and the reciprocal of that constant was placed before the integral sign. This procedure can be generalized for the exponential function as follows.

Suppose $f(x)$ is a function that can be expressed as $f(x) = e^{u(x)}g(x)$ dx. Calculate $du = u'(x)\,dx$. If possible find a constant k such that $kg(x)\,dx = du$. Then

$$\int e^{u(x)}g(x)\,dx = \frac{1}{k}\int e^{u(x)}kg(x)\,dx$$

$$= \frac{1}{k}\int e^u\,du = \frac{1}{k}\,e^u + C$$

Example 1 Consider

$$\int x\, e^{x^2}\, dx$$

Let $u(x) = x^2$ and $g(x) = x$. Hence $u'(x) = 2x$, and $du = 2x\, dx$. Therefore $k = 2$, since $2g(x)\, dx = du$. Therefore

$$\int xe^{x^2}\, dx = \frac{1}{2}\int e^{x^2} 2x\, dx$$

$$= \frac{1}{2}\int e^u\, du = \frac{1}{2}e^u + C = \frac{1}{2}e^{x^2} + C$$

Example 2 Consider

$$\int \frac{e^{1/x}\, dx}{x^2}$$

Let $u(x) = 1/x$, $g(x) = 1/x^2$. Hence $u'(x) = -1/x^2$ and $du = -dx/x^2$. Therefore $k = -1$, since $-g(x)\, dx = du$. Hence

$$\int \frac{e^{1/x}}{x^2}\, dx = -\int e^{1/x}\left(-\frac{dx}{x^2}\right)$$

$$= -\int e^u\, du = -e^u + C = -e^{1/x} + C$$

The formula $\int e^u\, du = e^u + C$, may also be written in the form

$$\int e^{f(x)} f'(x)\, dx = e^{f(x)} + C$$

Finally, using the previously stated formula

$$\frac{d}{dx}a^x = a^x \ln a$$

the following integration formula may easily be verified by differentiation:

$$\int a^x\, dx = \frac{a^x}{\ln a} + C$$

With regard to the integration of the logarithmic function, we have not as yet encountered any function whose derivative is $\ln x$, and hence we are not prepared at this point to determine an antiderivative of $\ln x$. We shall obtain it subsequently by using an integration technique called integration by parts.

However, the differentiation formula

$$\frac{d}{dx}(\ln x) = \frac{1}{x}$$

gives us the important integration formula:

$$\int \frac{1}{x}\, dx = \ln x + C \qquad x > 0$$

The restriction $x > 0$ is necessary in view of the fact that the domain of the function $\ln x$ is $(0,\infty)$. However, since

$$\frac{d}{dx}[\ln(-x)] = -\frac{1}{-x} = \frac{1}{x}$$

we can eliminate this restriction by writing

$$\int \frac{1}{x}\, dx = \ln|x| + C$$

It will be recalled that the integration formula

$$\int x^n\, dx = \frac{x^{n+1}}{x+1} + C$$

does not apply when $n = -1$. Since $1/x = x^{-1}$, the missing case has now been taken care of. Furthermore, if $u = f(x)$ and $du = f'(x)\, dx$, then we can write

$$\int \frac{du}{u} = \ln|u| + C$$

or

$$\int \frac{f'(x)}{f(x)}\, dx = \ln|f(x)| + C$$

Example 1 Find

$$\int \frac{x}{x^2 - 1}\, dx$$

Let $u = x^2 - 1$. Then $du = 2x\, dx$. Hence

$$\int \frac{x}{x^2-1}\, dx = \frac{1}{2}\int \frac{2x}{x^2-1}\, dx$$

$$= \frac{1}{2}\int \frac{du}{u} = \frac{1}{2}\ln|u| + C$$

$$= \frac{1}{2} \ln |x^2 - 1| + C$$

Example 2 Find

$$\int \frac{\sin x}{3 \cos x - 5}\, dx$$

Let $u = 3 \cos x - 5$. Then $du = -3 \sin x\, dx$. Hence

$$\int \frac{\sin x}{3 \cos x - 5}\, dx = -\frac{1}{3} \int \frac{-3 \sin x}{3 \cos x - 5}\, dx$$

$$= -\frac{1}{3} \int \frac{du}{u} = -\frac{1}{3} \ln |u| + C$$

$$= -\frac{1}{3} \ln |3 \cos x - 5| + C$$

Example 3 Find

$$\int \frac{dx}{x \ln x}$$

Let $u = \ln x$. Then $du = (1/x\, dx)$. Hence

$$\int \frac{dx}{x \ln x} = \int \frac{1/x}{\ln x}\, dx = \int \frac{du}{u}$$

$$= \ln |u| + C = \ln |\ln x| + C$$

Each of the differentiation formulas for the trigonometric functions (Section 5.8) implies, of course, a corresponding integration formula. For example, the formula

$$\frac{d}{dx}(\sin x) = \cos x$$

implies $\int \cos x\, dx = \sin x + C$. The formula

$$\frac{d}{dx}(\cos x) = -\sin x$$

implies $\int -\sin x\, dx = \cos x + C$, or, in the more usual form, $\int \sin x\, dx = -\cos x + C$. Going down the list of the trigonometric derivatives, we can easily deduce such formulas as

$$\int \sec^2 x\, dx = \tan x + C$$

$$\int \sec x \tan x \, dx = \sec x + C$$

To obtain a formula for $\int \tan x \, dx$ we make use of the identity

$$\tan x = \frac{\sin x}{\cos x}$$

and let $u = \cos x$. Then $du = -\sin x \, dx$. Hence

$$\int \tan x \, dx = \int \frac{\sin x}{\cos x} \, dx$$

$$= -\int \frac{-\sin x}{\cos x} \, dx = -\int \frac{du}{u}$$

$$= -\ln|u| + C = -\ln|\cos x| + C$$

However, since

$$|\cos x| = \frac{1}{|\sec x|}$$

$$\ln|\cos x| = \ln 1 - \ln|\sec x| = -\ln|\sec x|$$

Hence $$\int \tan x \, dx = \ln|\sec x| + C$$

By a similar procedure we find

$$\int \cot x = \ln|\sin x| + C$$

Finally, using some algebraic manipulation, we can derive the formulas

$$\int \sec x \, dx = \ln|\sec x + \tan x| + C$$

and $$\int \csc x \, dx = \ln|\csc x - \cot x| + C$$

As is the case with the integration formulas previously discussed, and in fact with any integration formula, each of the above trigonometric integration formulas can be generalized. For example if $u = f(x)$ and $du = f'(x) \, dx$, then

$$\int \sin u \, du = -\cos u + C$$

Example 4 Find $\int \cos 3x\, dx$.
Let $u = 3x$. Then $du = 3\, dx$. Hence

$$\int \cos 3x\, dx = \frac{1}{3}\int \cos 3x\,(3\, dx)$$

$$= \frac{1}{3}\int \cos u\, du = \frac{1}{3}\sin u + C$$

$$= \frac{1}{3}\sin 3x + C$$

Example 5 Find $\int x \tan (x^2)\, dx$.
Let $u = x^2$. Then $du = 2x\, dx$. Hence

$$\int x \tan (x^2)\, dx = \frac{1}{2}\int \tan (x^2)\,(2x\, dx)$$

$$= \frac{1}{2}\int \tan u\, du = \frac{1}{2}\ln|\sec u| + C$$

$$= \frac{1}{2}\ln|\sec (x^2)| + C$$

Example 6 Find $\int \tan^2 2x \sec^2 2x\, dx$.
Let $u = \tan 2x$. Then $du = 2 \sec^2 2x\, dx$. Hence

$$\int \tan^2 2x \sec^2 2x\, dx = \frac{1}{2}\int (\tan^2 2x)(2 \sec^2 2x\, dx)$$

$$= \frac{1}{2}\int u^2\, du = \frac{1}{2}\frac{u^3}{3} + C$$

$$= \frac{1}{2}\frac{\tan^3 2x}{3} + C$$

$$= \frac{1}{6}\tan^3 2x + C$$

Example 7 Find $\int \sec 3x\, dx$.
Let $u = 3x$. Then $du = 3dx$. Hence

$$\int \sec 3x\, dx = \frac{1}{3}\int \sec 3x(3\, dx)$$

$$= \frac{1}{3}\int \sec u\, du$$

$$= \frac{1}{3}\ln|\sec u + \tan u| + C$$

$$= \frac{1}{3} \ln|\sec 3x + \tan 3x| + C$$

The formulas for the derivatives of the inverse trigonometric functions also provide corresponding integration formulas. For example

$$\int \frac{dx}{\sqrt{1-x^2}} = \arcsin x + C$$

and

$$\int \frac{dx}{1+x^2} = \arctan x + C$$

However these formulas may readily be generalized as follows:

$$\int \frac{dx}{\sqrt{a^2-x^2}} = \arcsin \frac{x}{a} + C \qquad a > 0$$

$$\int \frac{dx}{a^2+x^2} = \frac{1}{a} \arctan \frac{x}{a} + C \qquad a > 0$$

These formulas may be verified by differentiation.

Example 8 Find

$$\int \frac{dx}{\sqrt{9-4x^2}}$$

Let $u = 2x$, $a = 3$. Then $du = 2\ dx$. Hence

$$\int \frac{dx}{\sqrt{9-4x^2}} = \frac{1}{2} \int \frac{2\ dx}{\sqrt{3^2-(2x)^2}}$$

$$= \frac{1}{2} \int \frac{du}{\sqrt{a^2-u^2}} = \frac{1}{2} \arcsin \frac{u}{a} + C$$

$$= \frac{1}{2} \arcsin \frac{2x}{3} + C$$

Example 9 Find

$$\int \frac{dx}{16+25x^2}$$

Let $u = 5x$, $a = 4$. Then $du = 5\ dx$. Hence

$$\int \frac{dx}{16+25x^2} = \frac{1}{5} \int \frac{5\ dx}{4^2+(5x)^2}$$

$$= \frac{1}{5} \int \frac{du}{a^2+u^2} = \frac{1}{5} \cdot \frac{1}{a} \arctan \frac{u}{a} + C$$

$$= \frac{1}{20} \arctan \frac{5x}{4} + C$$

The integration procedures employed here and in the preceding section involve a basic integration relationship that may be stated as follows:

$$\int f[g(x)]g'(x)\,dx = F[g(x)] + C$$

where $f(x) = F'(x)$, i.e., $F(x)$ is an antiderivative of $f(x)$. The truth of this formula may be demonstrated by differentiating the right-hand side, using the chain rule.

For example, consider $\int \cos 3x\,dx$. Here $f(x) = \cos x$, $F(x) = \sin x$, $g(x) = 3x$, $g'(x) = 3$. Then

$$\int \cos 3x\,dx = \frac{1}{3} \int \overbrace{\cos 3x}^{f[g(x)]}\,\overbrace{(3)}^{g'(x)}\,dx$$

$$= \frac{1}{3} \overbrace{\sin 3x}^{F[g(x)]} + C$$

Furthermore, if we let $u = g(x)$, $du = g'(x)\,dx$, the basic integration formula shown above assumes the form

$$\int f(u)\,du = F(u) + C$$

5.19 Integration by Parts

A very useful procedure called integration by parts enables us to find many integrals that could not be obtained by any of the preceding rules.

We begin by recalling the formula for differentiating a product of two functions:

$$\frac{d}{dx}[f(x)g(x)] = f(x)g'(x) + g(x)f'(x)$$

Hence

$$f(x)g'(x) = \frac{d}{dx}[f(x)g(x)] - g(x)f'(x)$$

Integrating each term, we have

$$\int f(x)g'(x)\,dx = f(x)g(x) - \int g(x)f'(x)\,dx \tag{1}$$

If we let $u = f(x)$, $v = g(x)$, $du = f'(x)\,dx$, and $dv = g'(x)\,dx$, then formula (1) may be written in the form

$$\int u\,dv = uv - \int v\,du \tag{2}$$

The successful use of this formula depends upon the proper choice of the factors $f(x)$ and $g'(x)$. Evidently, whatever choice is made for $g'(x)$, it must be a function that can be integrated.

Example 1 Find $xe^{-x}\,dx$.
Let $f(x) = x$, $g'(x) = e^{-x}$. Then $f'(x) = 1$, $g(x) = \int e^{-x}\,dx = -e^{-x}$. The constant of integration is superfluous at this point. Hence

$$\begin{aligned}\int xe^{-x}\,dx &= -xe^{-x} + \int e^{-x}\,dx \\ &= -xe^{-x} - e^{-x} + C \\ &= -e^{-x}(x+1) + C\end{aligned}$$

Note that another possible choice is $f(x) = e^{-x}$, $g'(x) = x$. However, it would not be successful since we would then have

$$f'(x) = -e^{-x}$$

$$g(x) = \int x\,dx = \frac{x^2}{2}$$

and in that case we would have

$$\int xe^{-x}\,dx = \frac{x^2}{2}e^{-x} + \int \frac{x^2}{2}e^{-x}\,dx$$

and the last integral would be more complicated than the original one.

In the last two examples, we shall use the notation of formula (2).

Example 2 Find $\int \ln x\,dx$.
Let $u = \ln x$, $dv = dx$. Then $du = (1/x)dx$ and $v = \int dx = x$. Hence

$$\begin{aligned}\int \ln x\,dx &= x\ln x - \int (x)\left(\frac{1}{x}\right)dx \\ &= x\ln x - \int dx = x\ln x - x + C \\ &= x(\ln x - 1) + C\end{aligned}$$

Example 3 Find $\int x^2 \cos x\, dx$.
Let $u = x^2$, $dv = \cos x\, dx$. Then du $= 2x\, dx$, and $v = \int \cos x\, dx = \sin x$. Hence

$$\int x^2 \cos x\, dx = x^2 \sin x - 2\int x \sin x\, dx$$

To obtain the last integral we again use the integration-by-parts procedure. Let $u = x$, $dv = \sin x\, dx$. Then $du = dx$, $v = \int \sin x\, dx = -\cos x$. Hence

$$\int x \sin x\, dx = -x \cos x + \int \cos x\, dx$$
$$= -x \cos x + \sin x$$

Therefore $$\int x^2 \cos x\, dx = x^2 \sin x - 2(-x \cos x + \sin x) + C$$
$$= x^2 \sin x + 2\, x \cos x - 2 \sin x + C$$

5.20 Areas Bounded by Curves

The calculation of areas bounded at least in part by curves has been discussed in Section 5.16. We now develop these matters somewhat further, with examples slightly more complicated than those previously considered.

Example 1 Find the area bounded by the curves $y = 4x^2$ and $x = 4y^2$.
Solution The area to be found is shaded in Figure 5.14. The curves intersect at the origin and at the point $(\frac{1}{4}, \frac{1}{4})$, as can be determined by solving the system consisting of the two equations.

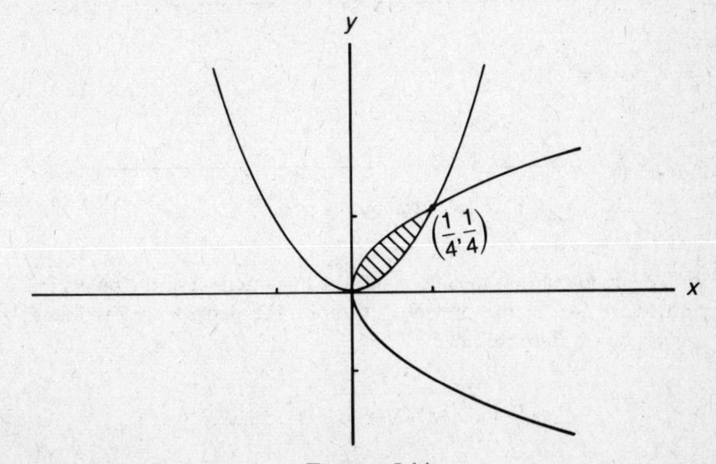

FIGURE 5.14

Evidently the desired area A can be found by subtracting the area A_1 bounded by the curve $y = 4x^2$, the x axis, and the line $x = \frac{1}{4}$ from the area A_2 bounded by the curve $x = 4y^2$ and the same lines. Hence

$$A = A_2 - A_1$$

$$= \int_0^{1/4} \frac{\sqrt{x}}{2}\, dx - \int_0^{1/4} 4x^2\, dx$$

where the first integral has been obtained by solving the equation $x = 4y^2$ for y and choosing the positive sign. Therefore

$$A = \int_0^{1/4} \left(\frac{x^{1/2}}{2} - 4x^2 \right) dx$$

$$= \left[\frac{x^{3/2}}{3} - \frac{4x^3}{3} \right]_0^{1/4}$$

$$= \frac{1}{24} - \frac{1}{48} = \frac{1}{48}$$

Example 2 Find the area bounded by the curve $y = 9 - x^2$ and the line $y = 5$.

Solution The area to be found is shaded in Figure 5.15.

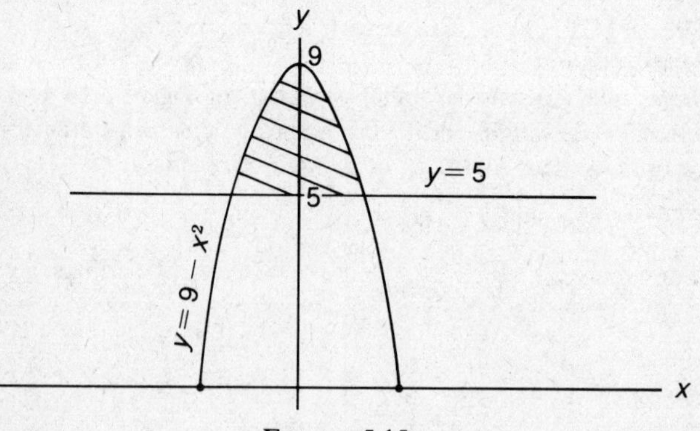

FIGURE 5.15

It is easily found that the curve and the line intersect at the points $(-2,5)$ and $(2,5)$. As in the previous example, the desired area is found by subtraction. Here we have

$$A = \int_{-2}^{2} (9 - x^2)\, dx - \int_{-2}^{2} 5\, dx$$

$$= \int_{-2}^{2} (4 - x^2)\, dx$$

$$= \left[4x - \frac{x^3}{3}\right]_{-2}^{2}$$

$$= \left(8 - \frac{8}{3}\right) - \left(-8 + \frac{8}{3}\right)$$

$$= \frac{32}{3}$$

Example 3 Find the area bounded by the curve $y = 1/x$, the x axis, and the lines $x = 1$ and $x = e$.

Solution The area to be found is shown in Figure 5.16.

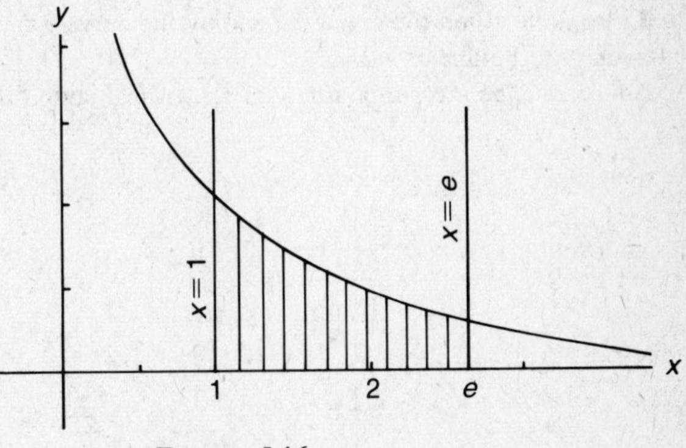

FIGURE 5.16

Hence $$A = \int_{1}^{e} \frac{1}{x}\, dx$$

$$= [\ln|x|]_1^e = \ln e - \ln 1 = 1$$

Example 4 Find the area bounded by the curve $y = \sin x$, and the x axis in the interval $[o, \pi]$

Solution The area to be found is shown in Figure 5.17.

Hence $$A = \int_{0}^{\pi} \sin x\, dx = [-\cos x]_0^{\pi}$$

$$= -\cos \pi + \cos 0$$

$$= 1 + 1 = 2$$

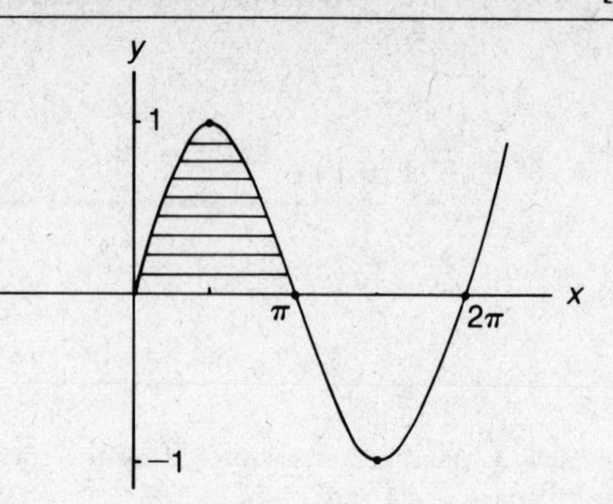

FIGURE 5.17

Example 5 Find the area bounded by the curve $y = e^{2x}$, the coordinate axes, and the line $x = 2$.

Solution The area to be found is shown in Figure 5.18.

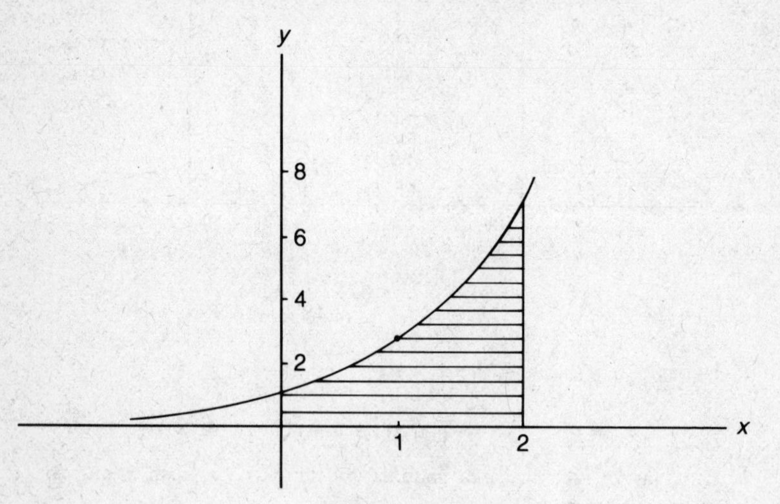

FIGURE 5.18

Hence
$$A = \int_0^2 e^{2x}\, dx$$
$$= [\tfrac{1}{2}e^{2x}]_0^2 = \tfrac{1}{2}e^4 - \tfrac{1}{2}e^0$$
$$= \frac{e^4 - 1}{2}$$

5.21 Linear Motion

In Section 5.11 we considered problems in which a particle moves on a coordinate line and in which the position of the particle was given as a function of the time. By differentiation we were then able to determine the velocity of the particle as a function of the time.

We now consider the reverse problem. It seems clear that if the velocity function is given, we should be able to obtain the position function by integration. However, a slight complication arises in view of the fact that the antiderivative of a given function is not unique. In addition to the velocity function, we need sufficient information to enable us to determine the constant of integration. Consider the following example.

Example 1 A particle moves along a coordinate line in such a way that its velocity at any time is given by $v(t) = 3t^2 - 5$. Its initial position is 3 units to the right of the origin. Find the position of the particle at $t = 4$.

Solution The statement regarding the initial position of the particle may be written as $x(0) = 3$. Then

$$x(t) = \left[\int (3t^2 - 5)\, dt\right]_{x(0)=3}$$
$$= (t^3 - 5t + C)_{x(0)=3}$$

Hence replacing t by 0 and $x(0)$ by 3, we have $C = 3$. Hence

$$x(t) = t^3 - 5t + 3$$
$$x(4) = 64 - 20 + 3 = 47$$

5.22 Improper Integrals

Integrals of the form

$$\int_a^\infty f(x)\, dx \qquad \int_{-\infty}^b f(x)\, dx \qquad \int_{-\infty}^\infty f(x)\, dx$$

are called **improper integrals.** They are defined as follows:

$$\int_a^\infty f(x)\, dx = \lim_{b\to\infty} \int_a^b f(x)\, dx \tag{1}$$

$$\int_{-\infty}^b f(x)\, dx = \lim_{a\to-\infty} \int_a^b f(x)\, dx \tag{2}$$

$$\int_{-\infty}^{\infty} f(x)\,dx = \int_{-\infty}^{a} f(x)\,dx + \int_{a}^{\infty} f(x)\,dx \tag{3}$$

In each case, if the limit exists, the integral is said to be convergent. Otherwise it is said to be divergent. Here are some examples.

Example 1 Find

$$\int_{1}^{\infty} \frac{1}{x^2}\,dx$$

$$\int_{1}^{\infty} \frac{1}{x^2}\,dx = \lim_{b\to\infty} \int_{1}^{b} \frac{1}{x^2}\,dx$$

$$= \lim_{b\to\infty} \left[-\frac{1}{x}\right]_1^b$$

$$= \lim_{b\to\infty} \left[-\frac{1}{b}+1\right] = 0+1 = 1$$

Example 2 Find

$$\int_{2}^{\infty} \frac{1}{x}\,dx$$

$$\int_{2}^{\infty} \frac{1}{x}\,dx = \lim_{b\to\infty} \int_{2}^{b} \frac{1}{x}\,dx$$

$$= \lim_{b\to\infty} \left[\ln|x|\right]_2^b$$

$$= \lim_{b\to\infty} (\ln b - \ln 2).$$

Since $\ln b$ increases without limit as b increases without limit, the given integral has no value; i.e., it is divergent.

Example 3 Find

$$\int_{-\infty}^{3} e^x\,dx$$

$$\int_{-\infty}^{3} e^x\,dx = \lim_{a\to-\infty} \int_{a}^{3} e^x\,dx$$

$$= \lim_{a\to-\infty} \left[e^x\right]_a^3$$

$$= \lim_{a\to-\infty} [e^3 - e^a] = e^3 - 0 = e^3$$

The fact that $\lim_{a\to-\infty} e^a = 0$, as indicated above, becomes clear when it is noted that $e^a = 1/e^{-a}$ and that $-a \to \infty$ as $a \to -\infty$.

A second type of improper integral occurs when the function to be integrated is unbounded in the interval of integration; i.e., the function increases or decreases beyond any finite limit.

Consider, for example,

$$\int_0^1 \frac{1}{x^{2/3}}\, dx$$

The function $1/x^{2/3}$ increases without limit as x approaches 0. Hence, the above integral is not defined in the ordinary sense. However, as in the case of the previously defined type of improper integral, this integral may be defined as a limit as shown below, where it is assumed that ϵ approaches 0 through positive values.

$$\begin{aligned}\int_0^1 \frac{1}{x^{2/3}}\, dx &= \lim_{\epsilon\to 0} \int_\epsilon^1 \frac{1}{x^{2/3}}\, dx \\ &= \lim_{\epsilon\to 0} \left[3x^{1/3} \right]_\epsilon^1 = \lim_{\epsilon\to 0} [3 - 3\epsilon^{1/3}] \\ &= 3 - 0 = 3\end{aligned}$$

5.23 Partial Derivatives

The idea of partial derivatives occurs in connection with functions of several variables, so we shall first explore this concept briefly.

We have already discussed (Section 2.1) the concept of a function as a set of ordered pairs, no two of which have the same first element. Such a function may be referred to as a function of one variable in the sense that the domain of the function consists of a set of individual numbers. We shall now extend this concept to functions of two or more variables.

A function of two variables may be defined as a correspondence between a set of ordered pairs called the domain and a set of real numbers called the range such that to each element of the domain there corresponds one and only one element of the range. A function of two variables may therefore be equivalently defined as a set of ordered triples, no two of which have the same first two elements. Thus, the set

$$\{(1,3,7),\ (1,4,9),\ (2,3,8),\ (2,4,10)\}$$

constitutes a function of two variables, whereas the set

$$\{(1,3,7), (1,3,9), (2,3,8), (2,4,10)\}$$

does not.

Just as in the single-variable case, functions of two variables are usually defined by means of algebraic expressions, for example

$$z = x^2 + 3y^2$$

Here, by assigning values to x and y, a set of ordered triples of the form (x,y,z) can be generated. For example, (1,1,4), (2,1,7), (2,2,16) are ordered triples belonging to the set that constitutes the function under discussion. Evidently the domain of this function is the set of all ordered pairs of real numbers, and the range is the set of all non-negative real numbers.

Using functional notation, the above function might be denoted by

$$f(x,y) = x^2 + 3y^2$$

Examples of functions of two variables are readily found in geometry and in physics. The volume v of a right circular cylinder is given by

$$v = \pi r^2 h$$

where r is the radius and h the altitude of the cylinder. One of Newton's laws of motion is expressed by the functional equation

$$F = kma$$

where F represents force, m is mass, a is acceleration, and k is a constant of proportionality whose value depends upon the units employed.

Functions of three or more variables may be defined by obvious extensions of the concept of a function of two variables. Thus a function of three variables may be defined as a set of ordered quadruples, no two of which have the same first three elements. An example of an algebraic representation of such a function is

$$w = x^2 - 3y + 4z^2$$

or

$$f(x,y,z) = x^2 - 3y + 4z^2$$

We now proceed to the concept of partial derivatives. A **partial derivative** of a function of several variables is a derivative obtained by regarding all of the variables except one as constants. To illustrate, let us return to the function $z = x^2 + 3y^2$. The partial derivative of z with respect to x is defined as the derivative obtained by differentiating with respect to x, considering y as a constant; it is denoted by the symbol $\partial z/\partial x$. Similarly, the partial derivative of z with respect to y is defined as the derivative obtained by differentiating with respect to y, considering x as a constant; it is denoted by the symbol $\partial z/\partial y$. Hence, for the function under consideration we have

$$\frac{\partial z}{\partial x} = 2x \qquad \frac{\partial z}{\partial y} = 6y$$

For a second example consider the function

$$z = x^2y + xy^3$$

In this case we obtain

$$\frac{\partial z}{\partial x} = 2xy + y^3$$

$$\frac{\partial z}{\partial y} = x^2 + 3xy^2$$

Finally, consider the function

$$z = e^{xy} + x \sin y$$

Here we obtain

$$\frac{\partial z}{\partial x} = ye^{xy} + \sin y$$

$$\frac{\partial z}{\partial y} = xe^{xy} + x \cos y$$

Higher-order partial derivatives are defined in a manner similar to higher-order derivatives for functions of one variable. Thus

$$\frac{\partial^2 z}{\partial x^2} = \frac{\partial}{\partial x}\left[\frac{\partial z}{\partial x}\right] \qquad \text{etc.}$$

There are also mixed higher-order partial derivatives, for example,

$$\frac{\partial^2 z}{\partial x\,\partial y} = \frac{\partial}{\partial x}\left[\frac{\partial z}{\partial y}\right]$$

$$\frac{\partial^2 z}{\partial y\,\partial x} = \frac{\partial}{\partial y}\left[\frac{\partial z}{\partial x}\right]$$

For the function, $z = e^{xy} + x \sin y$, using the partial derivatives already obtained, we can readily calculate

$$\frac{\partial^2 z}{\partial x^2} = \frac{\partial}{\partial x}(ye^{xy} + \sin y)$$

$$= y^2 e^{xy}$$

$$\frac{\partial^2 z}{\partial y^2} = \frac{\partial}{\partial y}(xe^{xy} + x \cos y)$$

$$= x^2 e^{xy} - x \sin y$$

$$\frac{\partial^2 z}{\partial y\, \partial x} = \frac{\partial}{\partial y}(ye^{xy} + \sin y)$$

$$= xye^{xy} + e^{xy} + \cos y$$

$$\frac{\partial^2 z}{\partial x\, \partial y} = \frac{\partial}{\partial x}(xe^{xy} + x \cos y)$$

$$= xye^{xy} + e^{xy} + \cos y$$

It will bc notcd that the product rule for derivatives was employed in obtaining the above mixed partial derivatives. It will also be noted that in this instance

$$\frac{\partial^2 z}{\partial x\, \partial y} = \frac{\partial^2 z}{\partial y \partial x}$$

i.e., the order in which the partial derivates is taken is immaterial. This state of affairs holds for any function $z = f(x,y)$ for which

$$z, \frac{\partial z}{\partial x}, \frac{\partial z}{\partial y}, \text{ and } \frac{\partial^2 z}{\partial x \partial y}$$

are continuous functions.

Among the applications of partial derivatives is their use in determining maximum or minimum values of functions of several variables. The procedure is somewhat similar to the one employed using ordinary derivatives for functions of one variable.

5.24 Differential Equations

We shall conclude our discussion of the calculus with a very brief discussion of the solution of differential equations.

A **differential equation** is simply an equation containing derivatives. A solution to a differential equation is a relationship among the variables that does not contain derivatives.

Anyone who has performed integrations has already solved simple differential equations. For example, consider the differential equation

$$\frac{dy}{dx} = 3x^2 + \cos x$$

A solution to this equation is readily obtained by integration as follows:

$$y = x^3 + \sin x + C$$

where C is an arbitrary constant. Consider next the differential equation

$$\frac{d^2y}{dx^2} = x + 4x^3$$

By repeated integration, we obtain

$$\frac{dy}{dx} = \frac{1}{2}x^2 + x^4 + C_1$$

and

$$y = \frac{1}{6}x^3 + \frac{1}{5}x^5 + C_1 x + C_2$$

Note, however, that the rather simple differential equation

$$\frac{dy}{dx} = y$$

cannot be solved by the method previously employed, since we would obtain

$$y = \int y\,dx$$

and we are unable to proceed since y is an unknown function of x. However, the equation may be solved by a procedure called **separation of variables,** i.e., by separating the terms involving x and dx from those involving y and dy. We can accomplish this in this case by multiplying the equation by dx/y to obtain $dy/y = dx$. If we now integrate both sides of the equation, we obtain

$$\ln y = x + C \qquad y > 0$$

It will be noted that the constant of integration has been placed on only the right-hand side of the equation, since a constant placed on the left could have been combined with the one on the right.

The above equation may be written in the form

$$y = e^{x+C}$$

But since $e^{x+C} = e^x e^C$, our solution may be written in the form

$$y = ke^x$$

where
$$k = e^C$$

For a second example of an equation that can be solved by separating the variables consider the equation

$$\frac{dy}{dx} = xe^y$$

Multiplying by $e^{-y}\,dx$, we obtain

$$e^{-y}\,dy = x\,dx$$

Integrating, we have

$$-e^{-y} = \frac{x^2}{2} + C$$

or
$$x^2 + 2e^{-y} = k$$

where
$$k = -2C$$

Consider next the equation

$$\frac{dy}{dx} + 3y = 0$$

This equation can be solved by separating the variables. However, we shall try a different approach. We shall propose, as a trial solution, a function of the form $y = e^{mx}$, where m is a constant. Then $dy/dx = me^{mx}$. Substituting in our equation, we have

$$me^{mx} + 3e^{mx} = 0$$

Hence
$$e^{mx}(m + 3) = 0$$

Since $e^{mx} \neq 0$, the only solution to the above equation is found by solving

$$m + 3 = 0$$

Hence
$$m = -3$$

Therefore, the function $y = e^{-3x}$ is a solution of the given differential equation. As a matter of fact, it is easily verified that $y = Ce^{-3x}$, where C is an arbitrary constant, is a solution.

Our next example is the equation

$$\frac{d^2y}{dx^2} + 3\frac{dy}{dx} - 10y = 0$$

Again we propose a trial solution of the form $y = e^{mx}$. Hence

$$\frac{dy}{dx} = me^{mx} \qquad \frac{d^2y}{dx^2} = m^2e^{mx}$$

Substituting in the differential equation, we obtain

$$m^2e^{mx} + 3me^{mx} - 10e^{mx} = 0$$

or
$$e^{mx}(m^2 + 3m - 10) = 0$$

Solving the quadratic equation

$$m^2 + 3m - 10 = 0$$

we obtain

$$(m + 5)(m - 2) = 0$$

or
$$m = -5 \qquad m = 2$$

Hence each of the functions $y = e^{-5x}$ and $y = e^{2x}$ is a solution of the given differential equation. It is readily verified that every function of the form $y = C_1e^{-5x} + C_2e^{2x}$, where C_1 and C_2 are arbitrary constants, is a solution.

The last two examples involved differential equations of a type called a **homogeneous linear differential equation.** It is called linear in the sense that the left-hand side consists of terms, each of which is of the first degree in y or one of its derivatives, and it is called homogeneous in the sense that the right-hand side is 0. For such an equation it can be shown that if each of $e^{m_1 y}$, $e^{m_2 y}$, . . . , $e^{m_k x}$ is a solution of the equation, then the linear combination

$$y = C_1 e^{m_1 x} + C_2 e^{m_2 x} + \cdots + C_k e^{m_k x}$$

where $C_1, C_2, \ldots C_k$ are arbitrary constants is also a solution.

We close this section with two examples illustrating the application of a very simple differential equation to problems involving the law of growth. The law of growth operates in a case in which the rate of change of the amount of some substance is, at all times, proportional to the amount present. This state of affairs can be represented by the differential equation

$$\frac{dA}{dt} = kA$$

where A is the amount of the substance, t represents time, and k is a constant. Examples of natural phenomena that are at least approximately subject to the law of growth are the growth of animal and human populations and the decomposition of radium. It is also exemplified in continuous compounding of interest on money placed on deposit in a savings institution.*

Example 1 The population of a certain city was 60,000 in 1970 and 66,000 in 1979. Assuming that the law of growth is applicable, find an expression for the population as a function of the time.

Solution If we denote the population by P, we have

$$\frac{dP}{dt} = kP$$

Separating the variables and integrating, we obtain

$$\frac{dP}{P} = k\,dt$$

$$\ln P = kt + C$$

$$P = e^{kt+C} = e^{kt}e^{C}$$

or

$$P = Ke^{kt}, \text{ where } K = e^{C}$$

Now, if we designate the year 1970 as time 0, we have, by the conditions of the problem,

(1) When $t = 0$, $P = 60{,}000$

(2) When $t = 9$, $P = 66{,}000$

Using (1) in the last equation above, we obtain

* Continuous compounding is the limiting case as the number of compounding periods per year increases without limit.

$$60{,}000 = Ke^0$$

or
$$K = 60{,}000$$

Hence our equation becomes

$$P = 60{,}000e^{kt}$$

Now using (2) in this equation, we obtain

$$66{,}000 = 60{,}000e^{9k}$$

From this we obtain

$$e^{9k} = \frac{66{,}000}{60{,}000} = 1.1$$

or
$$e^k = (1.1)^{1/9}$$

Hence, we have

$$P = 60{,}000\ [(1.1)^{1/9}]^t$$

or
$$P = 60{,}000\ (1.1)^{t/9}$$

This equation can be used to estimate the population in some future year. For example, for 1988, using $t = 18$, we would obtain

$$\begin{aligned} P &= (60{,}000)(1.1)^2 \\ &= 72{,}600 \end{aligned}$$

Example 2 If \$10,000 is invested at an annual rate of 6%, compounded continuously, express the amount of money present as a function of the time.

Solution In this case, if we designate the amount of money by A, we have

$$\frac{dA}{dt} = 0.06A \qquad \frac{dA}{A} = 0.06\,dt$$

$$\ln A = 0.06t + C \qquad A = e^{0.06t+C} = e^{0.06t}e^C$$

Hence $A = Ke^{0.06t}$

where $K = e^c$

since $A = 10{,}000$ when $t = 0$, $K = 10{,}000$

and we obtain $A = 10{,}000e^{0.06t}$

This equation can be used to calculate the amount present after t years. For example, for $t = 10$, we obtain

$$A = 10{,}000e^{0.6}$$

Using a table of the exponential function, we obtain

$$A = (10{,}000)(1.8221)$$
$$= \$18{,}221$$

Problems—Chapter 5

Derivatives of Algebraic Functions

1. Find $f'(x)$ for each of the following functions using the **definition** of the derivative:
   (a) $f(x) = 3x^2 - 2x$
   (b) $f(x) = \dfrac{2}{x^2}$
   (c) $f(x) = \dfrac{x}{x+2}$
   (d) $f(x) = mx + b$
2. Find dy/dx for each of the following functions, using the rules for differentiation:
   (a) $y = 7x^3 - 14x^2 + 5x - 9$
   (b) $y = 2x^{5/2} - 3x^{1/3}$
   (c) $y = \dfrac{7}{x} - \dfrac{4}{x^2}$
   (d) $y = \sqrt[3]{x} - \dfrac{1}{\sqrt{x}}$
   (e) $y = (3x - 5)^4$
   (f) $y = \sqrt{2x - 7}$
   (g) $y = \sqrt[3]{x^2 + 10}$
   (h) $y = \dfrac{1}{\sqrt{2x - 6}}$
   (i) $y = \left(1 - \dfrac{1}{x}\right)^5$
3. Find d^2y/dx^2 for each of the following functions:
   (a) $y = x^5 + 3x^4 - 2x^3 + 7x^2 + 8x - 10$
   (b) $y = (3x - 4)^5$
   (c) $y = \sqrt{x - 7}$
   (d) $y = \dfrac{2}{\sqrt{2x - 3}}$
4. Find dy/dx for each of the following functions:
   (a) $y = x^2\sqrt{4 - x}$
   (b) $y = \dfrac{x}{x^2 - 3}$
   (c) $y = \dfrac{x^2}{\sqrt{2x - 5}}$

Derivatives of Transcendental Functions

5. Find $f'(x)$ for each of the following functions:
   (a) $f(x) = \sin(x^3 - 4)$
   (b) $f(x) = \cos^2 3x$
   (c) $f(x) = \tan\sqrt{x}$
   (d) $f(x) = \sec x \tan x$

(e) $f(x) = \ln(x^3 - 4x)$

(f) $f(x) = \ln \dfrac{2x-3}{4x-5}$

(g) $f(x) = e^{\sin x}$

(h) $f(x) = e^x \cos 2x$

(i) $f(x) = \arcsin \sqrt{x}$

(j) $f(x) = \arccos (x^4)$

(k) $f(x) = e^x \arctan 2x$

(l) $f(x) = x^3 \cosh x$

Applications of the Derivatives

6. Find the equation of the tangent line to the graph of each of the following functions at the indicated point:
   (a) $f(x) = 3x^2 - 6$, (2,6)
   (b) $f(x) = \sin x$, $(\pi/6, \frac{1}{2})$
   (c) $f(x) = e^{3x}$, (0,1)
7. Find the extremes of each of the following functions:
   (a) $f(x) = \frac{1}{3}x^3 - x^2 - 8x + 5$
   (b) $f(x) = x^2 e^{-x}$
   (c) $f(x) = x^2 - 6x$, $x \in [1,6]$
8. Find two numbers whose sum is 10 and the sum of whose squares is a minimum.
9. A particle moves along a coordinate line in such a way that its position is given by $x(t) = t^3 - 3t + 4$, $t \geq 0$, where x is measured in feet and t in seconds.
   (a) What is the initial position of the particle?
   (b) What is its initial velocity?
   (c) During what time interval is the particle moving to the left?
   (d) What is the acceleration of the particle after 3 sec?
10. An object is released from a height of 384 ft with an upward velocity of 32 ft/sec.
    (a) When will the object strike the ground?
    (b) With what velocity will it strike the ground?
    (c) What will be the greatest height reached by the object?
11. A particle moves along the curve $y = \sqrt{x}$ in such a way that its abscissa (x coordinate) is increasing at the rate of 3 units/sec. At what rate is the ordinate (y coordinate) changing when the particle passes through the point (9,3)?
12. The edge of a cube is increasing at the rate of 5 cm/sec. At what rate is the volume changing when the edge is 4 cm in length?

13. Use derivatives to find the following limits:

(a) $\lim\limits_{x\to 0} \dfrac{x}{1-e^x}$

(b) $\lim\limits_{x\to 0} \dfrac{e^x - \cos x}{\sin x}$

(c) $\lim\limits_{x\to \infty} \dfrac{6x^2+4x-5}{2x^2-3x+4}$

Indefinite Integrals

14. Integrate each of the following functions:

(a) $\int (x^5+2x^3-5x+7)\,dx$

(b) $\int \sqrt{2x}\,dx$

(c) $\int \sqrt[3]{x}\,dx$

(d) $\int \left(\dfrac{1}{x^2}+\dfrac{2}{x^3}\right) dx$

(e) $\int \left(\dfrac{1}{\sqrt{x}}+\dfrac{2}{\sqrt[4]{x}}\right) dx$

(f) $\int (x-\sqrt{x})^2\,dx$

(g) $\int (2x-1)^3\,dx$

(h) $\int \left(\dfrac{x^2-x-2}{\sqrt{x}}\right) dx$

(i) $\int (2x-7)^8\,dx$

(j) $\int (5-3x)^4\,dx$

(k) $\int \sqrt{4x-7}\,dx$

(l) $\int \dfrac{dx}{\sqrt{3-x}}$

(m) $\int x\sqrt{x^2-5}\,dx$

(n) $\int \dfrac{x^2}{\sqrt[3]{x^3-7}}$

(o) $\int \dfrac{x-3}{(x^2-6x-1)^2}\,dx$

Particular Integrals

15. The slope of the tangent to a certain curve at any point is equal to the square of the abscissa (x coordinate) of the point. The curve passes through the point (3,4). Find the equation of the curve.

16. The amount A of a certain substance present at a given time varies with the time in such a way that $dA/dt = t^2 - \sqrt{t}$, where t is measured in seconds. If 20 units of the substance are present after 4 sec, find A as a function of t.

Indefinite Integrals Involving Transcendal Functions

17. Integrate each of the following functions:

(a) $\int e^{-4x}\,dx$

(b) $\int x^2 e^{-x^3}\,dx$

(c) $\int \frac{e^{\sqrt{x}}}{\sqrt{x}}\,dx$

(d) $\int \frac{dx}{3x-5}$

(e) $\int \frac{x}{4-x^2}\,dx$

(f) $\int \sin 4x\,dx$

(g) $\int \tan \frac{1}{2}x\,dx$

(h) $\int \sin^2 2x \cos 2x\,dx$

(i) $\int \frac{\cos x}{5 \sin x - 4}\,dx$

(j) $\int \frac{(\ln x)^3}{x}\,dx$

(k) $\int (1+\tan x)^3 \sec^2 x\,dx$

(l) $\int \frac{dx}{\sqrt{16-9x^2}}$

(m) $\int \frac{dx}{25+4x^2}$

Integration by Parts

18. Integrate each of the following functions:

(a) $\int xe^{-3x}\,dx$

(b) $\int x \sin x\,dx$

(c) $\int x^2 e^x\,dx$

(d) $\int x \ln x\,dx$

Definite Integrals

19. Evaluate each of the following:

(a) $\int_{-2}^{2} \sqrt{5-2x}\,dx$

(b) $\int_{0}^{3} \sqrt[3]{3x-1}\,dx$

(c) $\int_{4}^{6} (5-x)^{10}\,dx$

(d) $\int_{0}^{3} e^{2x}\,dx$

(e) $\int_{0}^{\pi/6} \cos 2x\,dx$

(f) $\int_{0}^{4} \frac{dx}{16+x^2}$

Applications of Integration

20. Find each of the following areas:

(a) The area bounded by the curves $y = x^2$ and $y = 8 - x^2$

(b) The area bounded by the curves $y = \sqrt{x}$ and $y = x^2$

(c) The area bounded by the curve $y = 8 - x^2$ and the line $y = 2x$

(d) The area bounded by the curve $y = e^{-x}$, the x axis, and the lines $x = -1$ and $x = 2$

21. A particle moves along a coordinate line in such a way that its velocity is given by $v(t) = 4 - t^3$. Its initial position is 2 units to the left of the origin. Find the position of the particle at $t = 3$.

Improper Integrals

22. Find the value of each of the following improper integrals, if it converges:

(a) $\int_2^\infty \frac{1}{x^3}\,dx$

(b) $\int_4^\infty \frac{1}{\sqrt{x}}\,dx$

(c) $\int_0^1 \frac{1}{x^{3/4}}\,dx$

(d) $\int_2^6 \frac{1}{\sqrt{x-2}}\,dx$

Partial Derivatives

23. Given the function $z = x^3y^2 + xy^4$, find:

(a) $\frac{\partial z}{\partial x}$

(b) $\frac{\partial z}{\partial y}$

(c) $\frac{\partial^2 z}{\partial x\,\partial y}$

24. Given the function $z = e^{x^2 y} + \sin xy$, find:

(a) $\frac{\partial z}{\partial x}$

(b) $\frac{\partial z}{\partial y}$

(c) $\frac{\partial^2 z}{\partial x\,\partial y}$

Differential Equations

25. Solve each of the following differential equations:

(a) $\frac{d^2y}{dx^2} = 3x - 4x^2$

(b) $\frac{dy}{dx} = 2y + 3$

(c) $\frac{d^2y}{dx^2} - 5\frac{dy}{dx} + 4y = 0$

26. The population of a certain city was 40,000 in 1970 and had fallen to 36,000 in 1975:

(a) Assuming that the law of growth is applicable, find an expression for the population P as a function of the time t.

(b) Estimate the population in 1980.

6

CONTINUOUS PROBABILITY

6.1 Continuous Sample Spaces and Events

In Section 4.2 a discrete sample space was defined as the finite or countably infinite set of all possible outcomes of an experiment. **Continuous sample spaces** correspond to experiments in which the number of possible outcomes is uncountably infinite. For example, consider the experiment of measuring the height of an individual in centimeters. Assuming the availability of sufficiently precise measuring instruments, the number of possible outcomes of the experiment is uncountably infinite. This is clear since even if it is known in advance that the height of the individual is between, say, 170 cm and 180 cm inclusive, the possible outcomes correspond to the points on a coordinate line in the interval [170,180], and these points cannot be put into one-to-one correspondence with the set of positive integers.

As in the case of the discrete sample space, an event is defined as a subset of the sample space, but in this case we restrict these subsets to those consisting of the union of finite or countably infinite sets of intervals. Referring to the previous example, we might refer to the event $\{[171,173] \cup [175,178]\}$.

6.2 Probability Distributions for Continuous Random Variables

The axioms used to define a probability function in Section 4.3 may be employed in the continuous case provided axiom (3) is extended as shown in Section 4.12. The notion of a random variable is also the same as in the discrete case. However, a random variable corresponding to a continuous sample space can, of course, assume an uncountably infinite number of values.

As was done in the discrete case, a random variable will be designed by a capital letter X, and a value of the random variable by the corresponding small letter x.

With the above stipulations in mind, we can now define a probability distribution for a continuous random variable as follows:

A probability distribution for a continuous random variable X is a function $f(x)$ for which

$$P(a \leq X \leq b) = \int_a^b f(x)\,dx$$

Hence it may be stated that if X is a continuous random variable whose distribution function is $f(x)$, then the probability that X will assume a value in the interval $[a,b]$ is the definite integral of $f(x)$ from a to b.

Since $\int_a^a f(x)\,dx = 0$, it follows that $P(X = a) = 0$. Hence

$$\begin{aligned} P(a \leq X \leq b) &= P(a < X \leq b) \\ &= P(a \leq X < b) = P(a < X < b) = \int_a^b f(x)\,dx \end{aligned}$$

It may seem puzzling that the probability is 0 that X will assume a specific value. First of all, it should be pointed out that whereas in the discrete case a zero probability corresponds to an impossible event, this is not true in the continuous case. Furthermore, no practical difficulties are encountered, since, due to the limitations of our measuring instruments, we are always concerned with the probability that X lies in an interval. For example, if a person's height is measured to the nearest tenth of a centimeter, then a measurement of 174.3 cm indicates that the true height lies in the interval (174.25,174.35).

A distribution function for a continuous random variable is commonly referred to as a **density function,** a term we shall use from this point on.

It is clear from the definition of the density function and from the axioms of a probability function that all density functions must possess the following properties:

(1) $f(x) \geq 0$

(2) $\displaystyle\int_{-\infty}^{\infty} f(x)\,dx = 1$

In connection with property (2) it should be noted that the possible values of a given random variable may lie in a finite interval. In this case, we simply define $f(x)$ to be 0 outside that interval, and (2) still holds.

For example, consider the density function defined as follows:

$$f(x) = \begin{cases} 2x & 0 \le x \le 1 \\ 0 & \text{elsewhere} \end{cases}$$

This function possesses property (1) since $2x \ge 0$, $0 \le x \le 1$. It also has property (2) since

$$\int_{-\infty}^{\infty} f(x)\,dx = \int_{-\infty}^{0} 0\,dx + \int_{0}^{1/2} 2x\,dx + \int_{1}^{\infty} 0\,dx$$

$$= 0 + [x^2]_0^1 + 0 = 1$$

Furthermore, using the same example, we obtain

$$P(0 < X < \tfrac{1}{2}) = \int_0^{1/2} 2x\,dx = [x^2]_0^{1/2} = \tfrac{1}{4}$$

and $$P(X > \tfrac{2}{3}) = \int_{2/3}^{1} 2x\,dx = [x^2]_{2/3}^{1} = 1 - \tfrac{4}{9} = \tfrac{5}{9}$$

It is readily seen that, in view of the definition of a density function, probabilities involving continuous random variables readily admit of a simple geometric interpretation. Thus,

$$P(0 < X < \tfrac{1}{2}) = \int_0^{1/2} 2x\,dx$$

is simply the area under the graph of $f(x) = 2x$ in the interval from $x = 0$ to $x = \frac{1}{2}$ as indicated in the shaded area in Figure 6.1. Also $P(X > \frac{2}{3})$ corresponds to the shaded area shown in Figure 6.2. It is clear from property (2) that for any density function the total area under the curve is equal to 1.

Example 1 Determine k for the following density function:

$$f(x) = \begin{cases} kx^2 & 1 \le x \le 2 \\ 0 & \text{elsewhere} \end{cases}$$

$$\int_1^2 kx^2\,dx = k\int_1^2 x^2\,dx = k\left[\frac{x^3}{3}\right]_1^2$$

$$= k(\tfrac{8}{3} - \tfrac{1}{3}) = \tfrac{7}{3}k = 1$$

Hence $k = \frac{3}{7}$, and the density function is

$$f(x) = \begin{cases} \frac{3}{7}x^2 & 1 \le x \le 2 \\ 0 & \text{elsewhere} \end{cases}$$

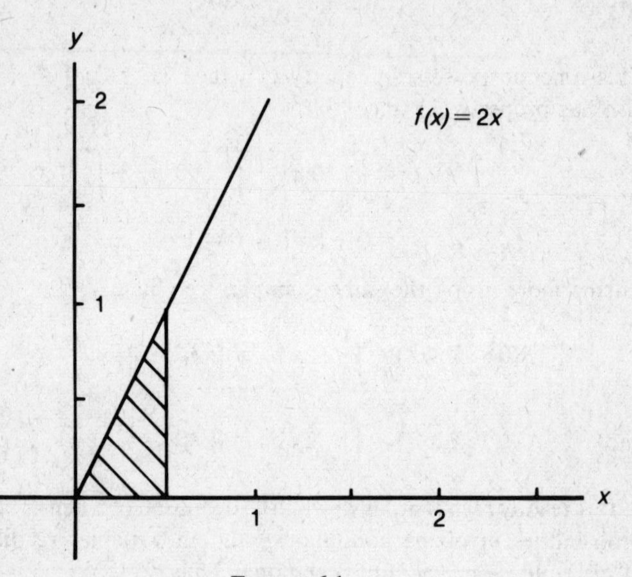

FIGURE 6.1

The **uniform density function** is defined as follows:

$$f(x) = \begin{cases} \dfrac{1}{b-a} & a \leq x \leq b \\ 0 & \text{elsewhere} \end{cases}$$

Example 2 If X has the uniform density function, where $a = 0$ and $b = 3$, find $P(X < 2)$.

$$P(X < 2) = \int_0^2 \tfrac{1}{3}\, dx = [\tfrac{1}{3}x]_0^2 = \tfrac{2}{3}$$

The **exponential density function** is defined as follows:

$$f(x) = \begin{cases} ae^{-ax} & a > 0,\ x \geq 0 \\ 0 & \text{elsewhere} \end{cases}$$

Example 3 If X has the exponential density function with $a = 2$, find $P(X < 1)$.

$$P(X < 1) = \int_0^1 2e^{-2x}\, dx$$

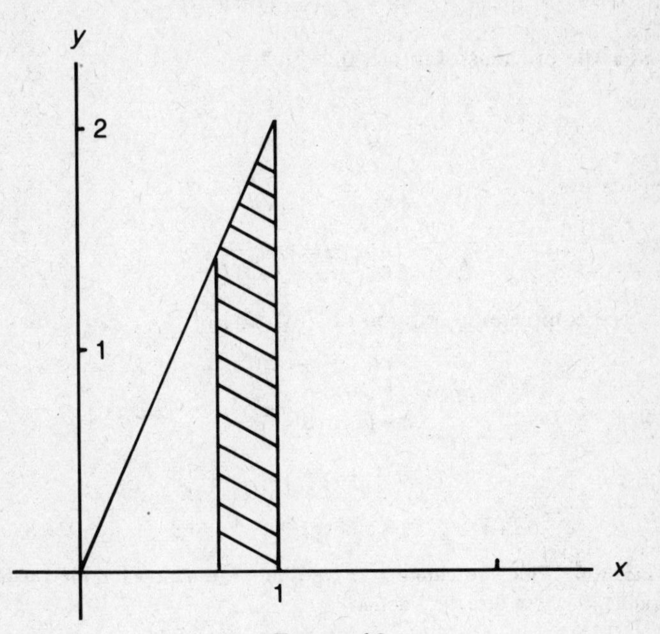

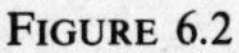
FIGURE 6.2

$$= [-e^{-2x}]_0^1$$

$$= 1 - e^{-2} \simeq 1 - 0.135$$

$$= 0.865$$

Closely related to the density function of a random variable is a function called the **cumulative distribution function.** If a random variable X has a density function $f(x)$, then clearly

$$P(X \leq a) = \int_{-\infty}^{a} f(x)\, dx$$

The function $F(x) = \int_{-\infty}^{x} f(t)\, dt$ is called the cumulative distribution function of the random variable X. It is also sometimes called simply the distribution function, which tends to confuse it with the probability distribution or density function.

Evidently the cumulative distribution function has the property that

$$F(a) = P(X \le a) = P(X < a)$$

Consider the previous example, in which

$$f(x) = \begin{cases} 2x & 0 \le x \le 1 \\ 0 & \text{elsewhere} \end{cases}$$

In this case, if $0 \le x \le 1$,

$$F(x) = \int_0^x 2t\,dt = [t^2]_0^x = x^2$$

Hence the complete description of $F(x)$ is

$$F(x) = \begin{cases} 0 & x < 0 \\ x^2 & 0 \le x \le 1 \\ 1 & x > 1 \end{cases}$$

Hence $$P(X < \tfrac{3}{4}) = F(\tfrac{3}{4}) = \tfrac{9}{16}$$

and $$P(X > \tfrac{1}{2}) = 1 - F(\tfrac{1}{2}) = 1 - \tfrac{1}{4} = \tfrac{3}{4}$$

Example 4 Find the cumulative distribution function for the random variable X whose density function is

$$f(x) = \begin{cases} 2e^{-2x} & x \ge 0 \\ 0 & \text{elsewhere} \end{cases}$$

For $x \ge 0$, $$F(x) = \int_0^x 2e^{-2t}\,dt$$

$$= [-e^{-2t}]_0^x = 1 - e^{-2x}$$

Hence the complete description of $F(x)$ is as follows:

$$F(x) = \begin{cases} 0 & x < 0 \\ 1 - e^{-2x} & x \ge 0 \end{cases}$$

6.3 The Expected Value and the Variance of Continuous Random Variables

The concepts of the expected value and the variance of a discrete random variable were discussed in Section 4.10. We now proceed to develop similar concepts applicable to continuous random variables.

For a continuous random variable X with density function $f(x)$ the expected value $E(X)$ is defined as

$$E(X) = \int_{-\infty}^{\infty} xf(x)\,dx$$

For example, consider once more the random variable X whose density function is

$$f(x) = \begin{cases} 2x & 0 \le x \le 1 \\ 0 & \text{elsewhere} \end{cases}$$

In this case

$$E(X) = \int_0^1 2x^2\,dx = \left[\frac{2x^3}{3}\right]_0^1 = \frac{2}{3}$$

For a second example, consider the random variable X whose density function is

$$f(x) = \begin{cases} 2e^{-2x} & x > 0 \\ 0 & \text{elsewhere} \end{cases}$$

In this case

$$E(X) = \int_0^{\infty} 2xe^{-2x}\,dx$$

$$= \lim_{b \to \infty} \int_0^b 2xe^{-2x}\,dx$$

Using integration by parts, we find

$$E(X) = \lim_{b \to \infty} [-xe^{-2x} - \tfrac{1}{2}e^{-2x}]_0^b$$

$$= \lim_{b \to \infty} (-be^{-2b} - \tfrac{1}{2}e^{-2b} + 0 + \tfrac{1}{2})$$

In evaluating the limits within the parentheses, note that the second term can be written as $-1/2e^{2b}$. The numerator of this fraction is constant, whereas as b increases without limit so does $2e^{2b}$. Hence the limit is 0. With regard to the first term, note that it can be written as $-b/e^{2b}$. The limit of this expression can be found using the procedure discussed in Section 5.13 and is seen to be 0. Hence $E(X) = \frac{1}{2}$.

The expected value of a continuous random variable may be interpreted, as in the discrete case, as the long-run average value of the variable. Hence it is often referred to as the mean and designated by the letter μ.

The variance of a continuous random variable X with density function $f(x)$ is denoted by the symbol σ^2 and is defined as follows:

$$\sigma^2 = E[(X-\mu)^2] = \int_{-\infty}^{\infty} (x-\mu)^2 f(x)\,dx$$

The square root of the variance is called the standard deviation and is denoted by σ.

Let us employ again the example of the random variable X whose density function is

$$f(x) = \begin{cases} 2x & 0 \le x \le 1 \\ 0 & \text{elsewhere} \end{cases}$$

We have already shown that $\mu = \frac{2}{3}$. Hence

$$\begin{aligned}\sigma^2 &= \int_0^1 \left(x - \frac{2}{3}\right)^2 2x\,dx \\ &= 2\int_0^1 \left(x^3 - \frac{4}{3}x^2 + \frac{4}{9}x\right) dx \\ &= 2\left[\frac{x^4}{4} - \frac{4x^3}{9} + \frac{2}{9}x^2\right]_0^1 \\ &= 2\left[\frac{1}{4} - \frac{4}{9} + \frac{2}{9}\right] = \frac{1}{18}\end{aligned}$$

As in the discrete case, the formula for the variance may be replaced by an equivalent expression that is usually simpler to use in calculation. It may be obtained as follows.

$$\begin{aligned}\sigma^2 &= \int_{-\infty}^{\infty} (x-\mu)^2 f(x)\,dx \\ &= \int_{-\infty}^{\infty} (x^2 - 2\mu x + \mu^2) f(x)\,dx \\ &= \int_{-\infty}^{\infty} x^2 f(x) - 2\mu \int_{-\infty}^{\infty} x f(x)\,dx + \mu^2 \int_{-\infty}^{\infty} f(x)\,dx\end{aligned}$$

But $\quad \int_{-\infty}^{\infty} x f(x)\,dx = \mu \quad$ and $\quad \int_{-\infty}^{\infty} f(x)\,dx = 1$

Hence
$$\sigma^2 = \int_{-\infty}^{\infty} x^2 f(x)\,dx - 2\mu^2 + \mu^2$$
$$= \int_{-\infty}^{\infty} x^2 f(x)\,dx - \mu^2$$

If this second formula is used in the last example, we have

$$\sigma^2 = \int_0^1 x^2(2x)\,dx - \left(\frac{2}{3}\right)^2$$
$$= 2\int_0^1 x^3\,dx - \frac{4}{9}$$
$$= 2\left[\frac{x^4}{4}\right]_0^1 - \frac{4}{9}$$
$$= \frac{1}{2} - \frac{4}{9} = \frac{1}{18}$$

This is the value obtained previously.

As a final example, let us calculate the mean and the variance of a random variable X having the uniform density function

$$f(x) = \begin{cases} \dfrac{1}{b-a} & a \le x \le b \\ 0 & \text{elsewhere} \end{cases}$$

Here
$$\mu = \frac{1}{b-a}\int_a^b x\,dx = \frac{1}{b-a}\left[\frac{x^2}{2}\right]_a^b$$
$$= \frac{1}{b-a}\left[\frac{b^2-a^2}{2}\right] = \frac{a+b}{2}$$

Also
$$\sigma^2 = \frac{1}{b-a}\int_a^b x^2\,dx - \left[\frac{a+b}{2}\right]^2$$
$$= \frac{1}{b-a}\left[\frac{x^3}{3}\right]_a^b - \frac{a^2+2ab+b^2}{4}$$
$$= \frac{1}{b-a}\left[\frac{b^3-a^3}{3}\right] - \frac{a^2+2ab+b^2}{4}$$
$$= \frac{b^2+ab+a^2}{3} - \frac{a^2+2ab+b^2}{4}$$

$$= \frac{4b^2 + 4ab + 4a^2 - 3a^2 - 6ab - 3b^2}{12}$$

$$= \frac{a^2 - 2ab + b^2}{12} = \frac{(a-b)^2}{12}$$

The significance of the variance of a continuous random variable is similar to that for the discrete case. Whereas the expected value provides a measure of average or central value, the variance provides a measure of dispersion, or spread, of the distribution. This point may be emphasized by considering the two following density functions:

$$f(x) = \begin{cases} \frac{3}{32}(4 - x^2) & -2 \leq x \leq 2 \\ 0 & \text{elsewhere} \end{cases} \tag{1}$$

$$g(x) = \begin{cases} \frac{1}{36}(9 - x^2) & -3 \leq x \leq 3 \\ 0 & \text{elsewhere} \end{cases} \tag{2}$$

We may easily verify that in each case $E(X) = 0$. However, it can also easily be verified that the variance of (1) is $\frac{4}{5}$, whereas the variance of (2) is $\frac{9}{5}$; i.e., the variance of (2) is over twice that of (1).

Clearly the variance serves to distinguish these two distributions

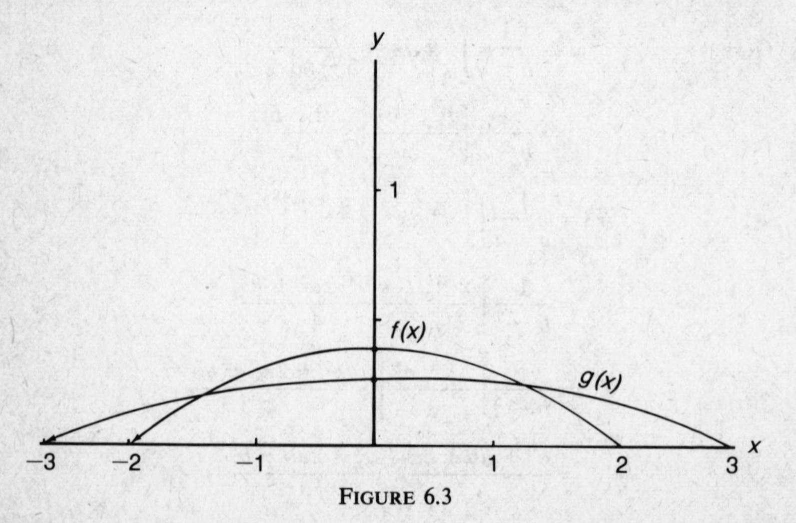

FIGURE 6.3

in an important way. The graphs of the two distributions are shown in Figure 6.3.

6.4 The Normal Distribution

The **normal density function,** often called the **normal distribution,** is given by

$$f(x) = \frac{1}{\sigma\sqrt{2\pi}}\, e^{-\frac{1}{2}\left(\frac{x-\mu}{\sigma}\right)^2} \qquad \infty < x < \infty$$

where all of the symbols except x represent constants. As previously indicated, e represents the base of the system of natural logarithms and it is approximately equal to 2.7183. An alternate way of writing e and its exponent is

$$\exp\left[-\frac{1}{2}\left(\frac{x-\mu}{\sigma}\right)^2\right]$$

The graph of the normal density function has essentially the same shape as the graph of the function $f(x) = e^{-x^2}$, or $f(x) = \exp(-x^2)$, which is shown in Figure 6.4.

The curve is seen to be bell-shaped and symmetrical with respect to the vertical axis.

The importance of the normal density function lies largely in the fact that many sets of observations found in the real world are approximately normally distributed. For example, distributions of heights, weights, and intelligence quotient's of human populations have this characteristic. The normal density function also plays a central

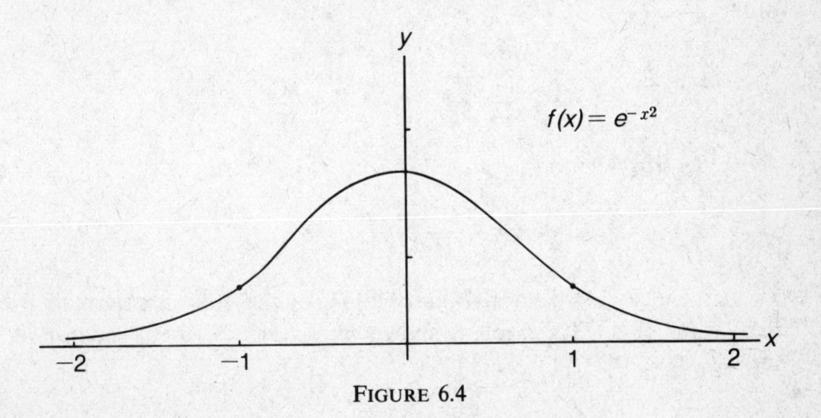

FIGURE 6.4

role in the development of the theory of probability and statistics.

The constants μ and σ appearing in the expression for the density function can be shown to be, in fact, the mean and the standard deviation of the distribution.

Furthermore, as is the case with all density functions,

$$\int_{-\infty}^{\infty} \frac{1}{\sigma\sqrt{2\pi}} \exp\left[-\frac{1}{2}\left(\frac{x-\mu}{\sigma}\right)^2\right] dx = 1$$

and $$P[a \leq X \leq b] = \int_a^b \frac{1}{\sigma\sqrt{2\pi}} \exp\left[-\frac{1}{2}\left(\frac{x-\mu}{\sigma}\right)^2\right] dx$$

At first glance it might appear that the calculation of probabilities involving normally distributed variables would involve the evaluation of rather complicated integrals. As a matter of fact the function $\exp(-x^2)$ is not integrable at all in the ordinary sense; i.e., there is no elementary function whose derivative is $\exp(-x^2)$. However, this presents no particular difficulties since definite integrals involving functions of this type can be readily approximated by various procedures such as the trapezoidal rule, which is discussed in all standard calculus texts.

As a matter of fact, probabilities involving normally distributed random variables can be obtained, using tables, as shown below.

The integral

$$\int_a^b \frac{1}{\sigma\sqrt{2\pi}} \exp\left[-\frac{1}{2}\left(\frac{x-\mu}{\sigma}\right)^2\right] dx$$

can be transformed using the substitution $z = (x - \mu)/\sigma$ to the form

$$\int_{(a-\mu)/\sigma}^{(b-\mu)/\sigma} \frac{1}{\sqrt{2\pi}} \exp\left(-\frac{1}{2} z^2\right) dz$$

The density function

$$f(z) = \frac{1}{\sqrt{2\pi}} \exp\left(-\frac{1}{2} z^2\right)$$

is called the **standard normal distribution,** or the **standard normal density function.** Its graph is shown in Figure 6.5, the standard normal curve.

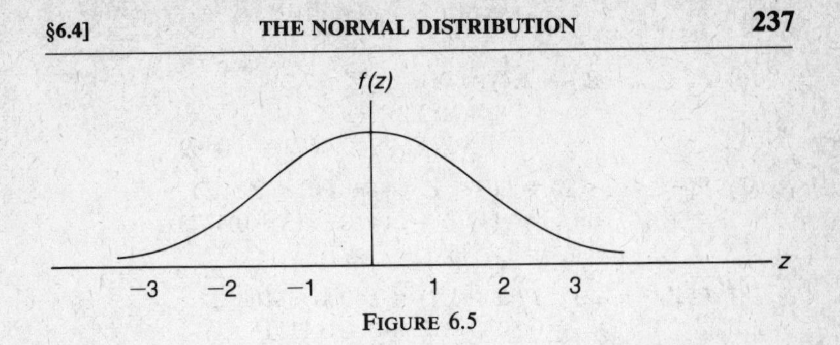

FIGURE 6.5

The total area under the curve in Figure 6.5 is 1. Since the curve is symmetrical with respect to the vertical axis, the areas from $-\infty$ to 0 and from 0 to ∞ are each equal to 0.5.

Furthermore, the random variable z, where $z = (X - \mu)/\sigma$, is called a **standard normal variable.** It is easily shown that it has mean 0 and variance 1. For example, if X is normally distributed with mean 2 and standard deviation 3, the standardized value corresponding to $X = 5$ is $Z = (5 - 2)/3 = 1$.

The function

$$A(z) = \int_0^z \frac{1}{\sqrt{2\pi}} \exp\left(-\frac{1}{2}t^2\right) dt$$

has been extensively tabulated. A table of this kind will be found in Appendix I. The table furnishes areas under the standard normal curve to the right of $z = 0$. Here are some illustrations of the calculation of probabilities obtained by using this table.

(1) $P(0 < Z < 1) = A(1) = 0.3413$

(2) $P(1 < Z < 2) = A(2) - A(1)$
$= 0.4773 - 0.3413 = 0.1360$

(3) $P(Z > 2.5) = 0.5000 - A(2.5)$
$= 0.5000 - 0.4938 = 0.0062$

(4) $P(Z < 2.0) = 0.5000 + A(2)$
$= 0.5000 + 0.4773 = 0.9773$

From the symmetry of the curve it can also be seen that

(5) $P(-1.5 < Z < 0) = P(0 < Z < 1.5)$
$= A(1.5) = 0.4332$

(6) $P(-2.2 < Z < -1.4) = P(1.4 < Z < 2.2)$
$= A(2.2) - A(1.4)$
$= 0.4861 - 0.4192 = 0.0669$

(7) $P(-1 < Z < 2) = P(0 < Z < 1) + P(0 < Z < 2)$
$= A(1) + A(2) = 0.3413 + 0.4773$
$= 0.8186$

(8) $P(Z < -1.2) = P(Z > 1.2) = 0.5000 - A(1.2)$
$= 0.5000 - 0.3849 = 0.1151$

The table of normal curve areas may also be used to find a value of Z corresponding to a specified probability. For example, suppose that it is desired to find the value z_0 such that $P(Z > z_0) = 0.20$. This is equivalent to $P(Z < z_0) = 0.80$, and since $P(Z < 0) = 0.50$, it is clear that $P(0 < Z < z_0) = 0.30$. Glancing over the table, we note that the area closest to 0.30 is 0.2995. This corresponds to $Z = 0.84$, and hence we conclude that $z_0 \simeq 0.84$.

For a second illustration of this type of calculation, suppose it is desired to find the value z_0 such that $P(-z_0 < Z < z_0) = 0.40$. Evidently in this case $P(0 < Z < z_0) = 0.20$, and we note that the tabular area closest to 0.20 is 0.1985, which corresponds to $Z = 0.52$. Hence $z_0 \simeq 0.52$.

In order to calculate probabilities involving normally distributed random variables with specified mean and standard deviation it is only necessary to obtain the Z values corresponding to the required X values and use the table as shown above.

Example 1 A random variable is normally distributed with mean 10 and standard deviation 2. Find

(a) $P(12 < X < 13)$

(b) $P(X > 14)$

(c) $P(9 < X < 11)$

(d) $P(X > 7)$

(e) $P(X < 13)$

Solution (a) If $X = 12$, then $Z = (12 - 10)/2 = 1$ and if $X = 13$, then $Z = (13 - 10)/2 = 1.5$. Therefore,

$$P(12 < X < 13) = P(1 < Z < 1.5) = A(1.5) - A(1)$$
$$= 0.4332 - 0.3413 = 0.0919$$

Solution (*b*) If $X = 14$, then $Z = (14 - 10)/2 = 2$.

$$\begin{aligned} P(X > 14) &= P(Z > 2) = 0.5000 - A(2) \\ &= 0.5000 - 0.4773 = 0.0227 \end{aligned}$$

Solution (*c*) If $X = 9$, then $Z = (9 - 10)/2 = -0.5$; if $X = 11$, then $Z = (11 - 10)/2 = 0.5$.

$$\begin{aligned} P(9 < X < 11) &= P(-0.5 < Z < 0.5) \\ &= 2A(0.5) = 2(0.1915) = 0.3830 \end{aligned}$$

Solution (*d*) If $X = 7$, then $Z = (7 - 10)/2 = -1.5$.

$$P(X > 7) = P(Z > -1.5) = 0.4332 + 0.5000 = 0.9332$$

Solution (*e*) $P(X < 13) = P(Z < 1.5) = 0.5000 + 0.4332 = 0.9332$

Example 2 A school that gives an entrance exam for admission wishes to admit approximately the upper 27% of the applicants taking the exam. Past experience has shown that the mean score on the exam is 80 and the standard deviation is 5. Assuming that a large group of applicants is to take the exam and that the school wishes to specify the passing grade in advance, what should the passing grade be?
Solution The value of Z that separates the lower 73% of the standard normal distribution from the upper 27% can be seen from the table to be approximately 0.61. Hence, if X represents an exam score, we must solve the equation

$$\frac{X - 80}{5} = 0.61$$

This gives $X = (5)(0.61) + 80 \simeq 83$ as the desired passing grade.

6.5 Approximating Binomial Probabilities Using the Normal Distribution

The binomial distribution for a discrete random variable has been discussed in Section 4.9. The computation of probabilities associated with binomially distributed random variables becomes very tedious if n is large. Fortunately, in cases of this kind, the time and labor involved in calculating such probabilities can be substantially reduced by using the normal distribution.

It can be shown, although the proof is beyond the scope of this book, that as the number of trials n becomes large, the binomial distribution approaches the normal distribution. Furthermore, the closer the value of the probability of success p is to $\frac{1}{2}$, the better the approximation will be for a given value of n.

For an illustration, consider the problem of calculating the probability that in 25 tosses of a balanced coin the number of heads will be from 11 to 13 inclusive.

The exact probability of this event can be obtained using the binomial distribution with $n = 25$ and $p = \frac{1}{2}$. The desired probability is the probability of 11, 12, or 13 successes, and the calculation can be shown symbolically as

$$P(X = 11, 12, \text{or } 13) = \sum_{x=11}^{13} \frac{25!}{x!(25-x)!}\left(\frac{1}{2}\right)^{25}$$

The result of this calculation can be viewed geometrically as the area of three rectangles of the histogram of the binomial distribution for which $n = 25$ and $p = \frac{1}{2}$. We intend to approximate this area by calculating the corresponding area under the normal curve for the distribution having the same mean and standard deviation as the given binomial distribution. See Figure 6.6.

In Section 4.11, we found that for a binomial distribution the mean and the variance are given by

$$\mu = np \qquad \sigma^2 = np(1-p)$$

Hence in the present case we have

$$\mu = (25)(\tfrac{1}{2}) = 12.5$$
$$\sigma^2 = (25)(\tfrac{1}{2})(\tfrac{1}{2}) = 6.25$$

Hence the standard deviation is

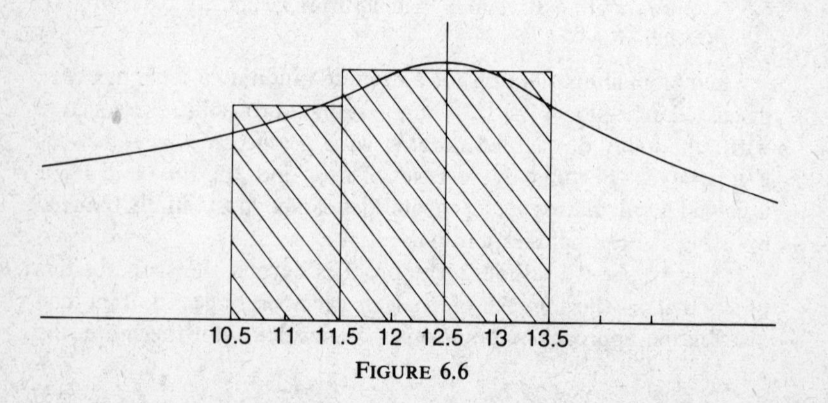

FIGURE 6.6

$$\sigma = \sqrt{6.25} = 2.50$$

Therefore the standard normal variable for this distribution is

$$Z = \frac{X - 12.5}{2.50}$$

To obtain the values of X that will provide the limits over which the standard normal density function is to be integrated, note that in Figure 6.6 the rectangles whose area we wish to approximate extend from 10.5 to 13.5 on the horizontal axis. We therefore compute:

$$Z_1 = \frac{10.5 - 12.5}{2.50} \qquad Z_2 = \frac{13.5 - 12.5}{2.50}$$

$$= -0.80 \qquad\qquad = 0.40$$

Hence the desired probability is

$$A(0.80) + A(0.40) = 0.2881 + 0.1554 = 0.4435$$

The normal approximation to the binomial may be expected to be reasonably accurate when $np > 5$ and $n(1 - p) > 5$. Note that the previous example amply meets this test since $np = n(1 - p) = (25)(\frac{1}{2}) = 12.5$.

Problems—Chapter 6

1. Determine k for the following density function:

$$f(x) = \begin{cases} kx^3 & 0 \leq x \leq 1 \\ 0 & \text{elsewhere} \end{cases}$$

2. If x has the density function

$$f(x) = \begin{cases} 3x^2 & 0 \leq x \leq 1 \\ 0 & \text{elsewhere} \end{cases}$$

find:
(a) $P(x < \frac{1}{3})$
(b) $P(\frac{1}{4} < x < \frac{1}{2})$

3. If x has the density function

$$f(x) = \begin{cases} \frac{1}{2}\, e^{-x/2} & x \geq 0 \\ 0 & \text{elsewhere} \end{cases}$$

find $P(x < 1)$

4. If x has the density function

$$f(x) = \begin{cases} \frac{1}{2}x & 0 < x < 2 \\ 0 & \text{elsewhere} \end{cases}$$

   (a) Find the cumulative distribution function $F(x)$.
   (b) Use the result of (*a*) to find $P(x < \frac{3}{2})$.
5. Find the expected value and the variance of the random variable defined in Problem 4.
6. A random variable x is normally distributed with mean 30 and standard deviation 5. Find:
   (a) $P(35 < x < 40)$
   (b) $P(x > 37)$
   (c) $P(25 < x < 38)$
   (d) $P(x < 22)$
   (e) $P(x < 39)$
7. The scores on an aptitude test are normally distributed with mean 84 and standard deviation 4. What score separates the upper 30% from the lower 70%.
8. Use the normal approximation to the binomial distribution to calculate the probability that in 36 tosses of a balanced coin the number of heads will be between 15 and 21 inclusive.

7

STATISTICAL INFERENCE

7.1 Introduction

In this chapter we shall apply the principles of discrete and continuous probability theory to the problems of statistical inference. **Statistical inference** refers to the process of reaching conclusions about the characteristics of a large set of objects called the **population** by using information obtained from a relatively small subset of the population called the **sample.** The problems of statistical inference are usually thought of as belonging to one of two types, estimation and tests of hypotheses. These terms will now be briefly discussed.

The term **estimation** is, of course, self-explanatory. An estimate is made of some characteristic of a population on the basis of some observed characteristic of the sample. Suppose, for example, that an urn contains a very large number of balls, each of which is known to be either blue or white, but that the proportion of blue and white balls is unknown. In order to estimate the proportion, the following procedure is employed. The balls in the urn, which constitute the population, are thoroughly mixed and then a sample of 10 balls is drawn from the urn. It is observed that 6 of these balls are blue. On this basis, it is estimated that three-fifths of the balls in the urn are blue; i.e., the sample proportion is used as an estimate of the population proportion.

Now, the above procedure may appear to be a rather obvious one, hardly requiring any knowledge of probability. However, as we shall presently see, the theory of probability enables us not only to make the above estimate, but also to include a measure of the reliability of the estimate. Clearly this is very important, since an estimate is of little use unless accompanied by some indication of the degree of confidence that can be placed in it.

To illustrate the **testing of hypotheses,** let us suppose that an assertion has been made that the proportion of blue balls in the urn is $\frac{4}{5}$. This assertion is called a hypothesis. A test of this hypothesis is a procedure for arriving at a decision either to accept the hypothesis as true or to reject it as false. Suppose, as before, that a sample of 10 balls is observed to contain 6 blue balls. The sample proportion $\frac{3}{5}$ differs from the asserted value $\frac{4}{5}$, but it is, of course, quite conceivable that this sample value could have been obtained even though the hypothesis is true. It is also clear that the more a sample proportion differs from the true population proportion, the less likely it is to occur. It is therefore necessary in order to reach a decision in this case to assess the probability of obtaining a sample value that differs from the asserted population value by as much as was the case in this instance and to arrive at a decision on that basis. The details of this procedure will be presently discussed more fully.

7.2 Descriptive Statistical Measures

Before continuing our discussion of statistical inference, we shall now conduct a brief discussion of some well-known descriptive statistical measures. As the term implies, these quantities provide a numerical description of statistical data, and a number of them are extensively used in statistical inference. Furthermore, each of these measures, as obtained from a sample, has a counterpart for the population. The sample measure is called a **statistic,** whereas the corresponding population measure is called a **parameter.**

We shall discuss first a group of statistics called **measures of central tendency** since they provide a central or representative value for the entire sample. The best known measure of central tendency is the **arithmetic mean,** which is denoted by the symbol $\bar{x}$. If a sample consists of the n values $x_1, x_2, \ldots, x_n$, then by definition

$$\bar{x} = \frac{\sum_{i=1}^{n} x_i}{n}$$

It is readily seen that the arithmetic mean is simply the quantity commonly referred to as the average value, since it is obtained by adding the values in the sample and dividing by the number in the sample. The number in the sample is often called the **sample size.**

The arithmetic mean is often referred to simply as the mean. However, there are two other types of means. The geometric mean is defined as follows:

$$\text{G.M.} = \sqrt[n]{x_1 \cdot x \cdot x_2 \cdot \cdot \cdot x_n}$$

Hence, the **geometric mean** is the nth root of the product of the sample values. The **harmonic mean** is defined as follows:

$$\text{H.M.} = \frac{1}{\frac{1}{n}\sum_{i=1}^{n}\frac{1}{x_i}}$$

Hence, the harmonic mean is the reciprocal of the arithmetic mean of the reciprocals of the sample values.

All of these means may properly be called averages. Although the appropriate average in a given situation is most often the arithmetic mean, this is not always the case, as is demonstrated in the following example.

Example 1 A journey of 60 mi is made at a constant speed of 30 mi/h. The return trip is made at a constant speed of 20 mi/h. What is the average speed for the trip?

Most people would be inclined to give 25 mi/h, the arithmetic mean, as the answer to the above problem. But let us see if this is correct. The data are shown in the following table.

	Distance	*Speed*	*Time*
Going	60 mi	30 mi/h	2 h
Returning	60	20	3
Totals	120		5

$$\text{Average speed} = \frac{120 \text{ mi}}{5 \text{ h}} = 24 \text{ mi/h}$$

Thus the average speed is not the arithmetic mean. In fact, it is the harmonic mean. The harmonic mean of 20 and 30 is given by

$$\text{H.M.} = \frac{1}{\frac{1}{2}\left(\frac{1}{20}+\frac{1}{30}\right)} = \frac{1}{\frac{1}{2}\left(\frac{5}{60}\right)} = \frac{120}{5} = 24$$

Another well-known measure of central tendency is the median. The **median** of a sample is the middle value when the sample values are arranged in order of magnitude. For example, consider the sample: 5, 4, 2, 7, 9. When these values are arranged in order of magnitude, we have 2, 4, 5, 7, 9. Hence the median is 5. If the number of sample values is even, there is, of course, no middle value. In this case, the median is defined as the number halfway between the two middle values.

The median is useful as a descriptive measure because it is not affected by extremely large or extremely small values. For example, consider the samples 3, 5, 6, 9, 10 and 3, 5, 6, 9, 50. The median of both samples is 6, but the mean of the second sample is considerably larger than that of the first. For this reason the median is sometimes a more representative value than the mean. For example, in connection with salary data, the median salary is often given rather than the mean, since it is not affected by a few large salaries that are not typical of the group as a whole.

We discuss next a group of statistics known as measures of dispersion. Two sets of data with the same mean may differ to a considerable extent in the **dispersion,** or spread, of the observations on either side of the mean. This idea has already been explored in Chapter 4 in connection with probability distributions.

The quantity $x_i - \bar{x}$ is the deviation of a specified observation from the mean, and it seems logical that these deviations would somehow be involved in the calculation of a measure of dispersion. However, it is not feasible to use the sum of these deviations since it can easily be shown that for any sample

$$\sum_{i=1}^{n} (x_i - \bar{x}) = 0$$

This difficulty can be avoided in two ways.

First, we can employ the sum of the absolute values of the deviations, which leads to the definition of the **mean deviation:**

$$\text{M.D.} = \frac{\sum_{i=1}^{n} |x_i - \bar{x}|}{n}$$

For example, consider the sample

$$2, 4, 3, 9, 7$$

We readily obtain $\bar{x} = 5$ and the absolute values of the deviations from the mean shown below:

x	2	4	3	9	7
$\lvert x-\bar{x}\rvert$	3	1	2	4	2

Hence $$\text{M.D.} = \frac{3+1+2+4+2}{5} = 2.4$$

A second approach to the construction of a measure of dispersion is to employ the squares of the deviations from the mean. This leads to the definition of a widely used statistic called the **variance.** The sample variance is usually denoted by the symbol s^2 and is defined as follows:

$$s^2 = \frac{\sum_{i=1}^{n} (x_i - \bar{x})^2}{n-1}$$

The presence of the factor $n-1$ instead of n in the denominator of the formula for the variance may seem strange at first sight. However, it is explained by the fact that in a certain sense the sample variance, as defined above, provides a better estimate of the population variance than would be the case if the factor n were used in the denominator.

Closely related to the variance is a measure called the **standard deviation,** which is defined as the non-negative square root of the variance and is denoted by the symbol s. Hence

$$s = \sqrt{\frac{\sum_{i=1}^{n} (x_i - \bar{x})^2}{n-1}}$$

For the sample employed in connection with the mean deviation the variance is calculated as follows:

$$s^2 = \frac{(2-5)^2 + (4-5)^2 + (3-5)^2 + (9-5)^2 + (7-5)^2}{4}$$

$$= \frac{9+1+4+16+4}{4} = 8.50$$

Hence $s = \sqrt{8.50} \approx 2.92$

7.3 The Distribution of Sample Statistics

Let us suppose that repeated random samples of size n are drawn from a population and that some statistic is calculated for each of the samples. The distribution of this statistic constitutes what is known as a **sampling distribution.**

It can be shown that if the population is normally distributed, then the sampling distribution of the sample mean is also normal. Furthermore, the mean of the distribution is equal to the population mean. The mean of a population is commonly denoted by the letter μ. The fact just stated may be written symbolically as*

$$\mu_{\bar{X}} = \mu_X \tag{1}$$

It is also true that the standard deviation of the distribution of sample means is equal to the standard deviation of the population divided by the square root of the sample size. The standard deviation of a population is usually denoted by the letter σ, and the statement just made may be written symbolically as

$$\sigma_{\bar{X}} = \frac{\sigma_X}{\sqrt{n}} \tag{2}$$

Equations (1) and (2) show that the population of means of samples of a fixed size drawn from a normally distributed population is centered about the same value as the original population but that its dispersion is smaller. Furthermore, the larger the sample size is, the smaller the dispersion of the sample means will be.

Since the means of samples drawn from normal populations are themselves normally distributed, it follows that the variable

$$Z = \frac{\bar{X} - \mu}{\sigma/\sqrt{n}}$$

has the standard normal distribution. This distribution was discussed in Section 6.4. Hence, probabilities involving the value of the sample mean can be calculated using the table of areas under the standard normal curve.

Example 1 A sample of size 16 is to be drawn from a normally distributed population with mean 50 and standard deviation 8. What is the probability that the mean of the sample will be between 48 and 52?

Using our previous formulas, we have

* In accordance with the previous notation, $\bar{X}$ is a random variable and $\bar{x}$ is a particular value of the random variable.

$$\mu_{\bar{X}} = 50$$

$$\sigma_{\bar{X}} = \frac{8}{\sqrt{16}} = 2$$

$$z_1 = \frac{48 - 50}{2} = -1$$

$$z_2 = \frac{52 - 50}{2} = 1$$

Hence
$$P(48 < \bar{X} < 52) = P(-1 < Z < 1)$$
$$= 2\,P(0 < Z < 1) = 2(0.34) = 0.68$$

It has been stated that the means of samples drawn from normal populations are also normally distributed. A remarkable theorem known as the **central limit theorem** ensures that the distribution of the sample mean tends to normality with increasing sample size regardless of the distribution of the population. The extremely important consequence of this fact is that, for sufficiently large samples, normal curve methods may be applied to any population whatsoever.

But when is a sample sufficiently large? Experience has shown that samples of size 30 or more are usually large enough to ensure that the sample mean is approximately normally distributed within limits of reasonable accuracy.

Example 2 A sample of size 64 is drawn from a population of unknown distribution, with mean 40 and standard deviation 4. What is the probability that the sample mean will exceed 41?

In this case we have

$$\mu_{\bar{X}} = 40$$

$$\sigma_{\bar{X}} = \frac{4}{\sqrt{64}} = \frac{1}{2}$$

$$z = \frac{41 - 40}{1/2} = 2$$

Hence
$$P(\bar{X} > 41) \simeq P(Z > 2) = 0.50 - 0.48 = 0.02$$

7.4 Estimating a Population Mean

A common problem in estimation is that of estimating the mean of a population using information furnished by a random sample drawn from the population. The sample mean seems to be the logical statistic to use in making this estimate although alternative procedures are available. The fact that the mean of the distribution of sample means is equal to the population mean recommends it as an estimator of the corresponding population parameter. The possession of this property qualifies the sample mean as an unbiased estimator. In general, a sample statistic is said to be an **unbiased estimator** of a population parameter if the mean of its sampling distribution is equal to the population parameter. The possession of this property ensures that the estimate will be correct *on the average.*

The fact that an estimate of the population mean is unbiased does not, of course, ensure that it is equal to or even close to the true value. However, a measure of the degree of confidence that can be placed in the estimate is provided by what is called a **confidence interval.**

We have seen that since the sample mean is, for large samples, at least approximately normally distributed, the normal distribution can be used to determine probabilities involving the value of the sample mean. For example, since the probability is approximately 0.95 that $-2 < Z < 2$, it can be stated with the same probability that

$$-2 < \frac{\bar{X} - \mu}{\sigma/\sqrt{n}} < 2$$

But this inequality is equivalent to

$$-2\frac{\sigma}{\sqrt{n}} < \bar{X} - \mu < 2\frac{\sigma}{\sqrt{n}}$$

or

$$-\bar{X} - \frac{2\sigma}{\sqrt{n}} < -\mu < -\bar{X} + 2\frac{\sigma}{\sqrt{n}}$$

Multiplying throughout by -1, we have

$$\bar{X} + 2\frac{\sigma}{\sqrt{n}} > \mu > \bar{X} - 2\frac{\sigma}{\sqrt{n}}$$

Evidently this inequality can be written in the reverse order as

$$\bar{X} - 2\frac{\sigma}{\sqrt{n}} < \mu < \bar{X} + 2\frac{\sigma}{\sqrt{n}}$$

The interval $\left(\bar{X} - 2\frac{\sigma}{\sqrt{n}}, \bar{X} + 2\frac{\sigma}{\sqrt{n}}\right)$

is called a 95% confidence interval for the population mean. A more exact interval is

$$\left(\bar{X} - 1.96\frac{\sigma}{\sqrt{n}}, \bar{X} + 1.96\frac{\sigma}{\sqrt{n}}\right)$$

Example 1 A sample of size 36 is drawn from a population whose standard deviation is known to be 6. If the sample mean is found to be 25, determine a 95% confidence interval for the population mean.

Solution Using the information given above, we have for the desired interval

$$25 - 2\left(\frac{6}{\sqrt{36}}\right) < \mu < 25 + 2\left(\frac{6}{\sqrt{36}}\right)$$

or

$$23 < \mu < 27$$

The interval [23,27] is a 95% confidence interval for the population mean in the following sense. If samples of size 36 were to be repeatedly drawn from this population, and, if in each case a confidence interval were to be calculated in the same way as has just been done, then in the long run 95% of these intervals would contain the true value of the population mean.

The procedure described above can readily be modified to obtain intervals with different levels of confidence. For example, since $P(-2.58 < Z < 2.58) \simeq 0.99$, it follows that a 99% confidence interval is given by

$$\bar{X} - 2.58\frac{\sigma}{\sqrt{n}} < \mu < \bar{X} + 2.58\frac{\sigma}{\sqrt{n}}$$

It will be noted that the calculation of a confidence interval for the population mean requires a knowledge of the population standard deviation. However, if this parameter is unknown, and if the sample is reasonably large, the sample standard deviation may be used as an estimate of the population value in carrying out the calculation.

7.5 Estimating a Population Proportion

The confidence interval technique employed in the previous section can be readily employed to estimate the proportion of a population falling in one of two possible categories. For example, it might be desired to estimate the proportion of defectives in a population in which each member is classified either as defective or nondefective.

The population proportion, which is usually designated by p, is estimated by drawing a random sample and calculating the sample proportion p'.* It can be shown that the mean of the sampling distribution of p' is p, and hence p' is an unbiased estimator of p. Furthermore, the standard deviation of the distribution of p' is

$$\sigma_{p'} = \sqrt{\frac{p(1-p)}{n}}$$

Finally, for large samples, the distribution of p' is approximately normal.

In view of the facts stated above, confidence intervals for the population proportion p can be constructed in a manner similar to that employed in estimating the mean μ. For example, a 95% confidence interval for p would assume the form

$$p' - 2\sqrt{\frac{p(1-p)}{n}} < p < p + 2\sqrt{\frac{p(1-p)}{n}}$$

It will be noted that the first and third members of this inequality involve p, the parameter to be estimated. However, this difficulty is surmounted by using

$$\sqrt{\frac{p'(1-p')}{n}}$$

to estimate

$$\sqrt{\frac{p(1-p)}{n}}$$

This substitution is justified by the fact that the change in the value of the expression $\sqrt{p(1-p)}$ is very small for small changes in the value of p. It can be shown, using the calculus, that the maximum value of $p(1-p)$ is $\frac{1}{4}$. The reader may verify this as an exercise.

Example 1 It is desired to estimate the proportion of Democrats in a population of voters. A random sample of size 100 is drawn, and it is observed that 64 of the sample are Democrats. Construct a 99% confidence interval for the proportion of Democrats in the population.

Solution In this case the sample proportion is

* Note that in this case the random variable is denoted by p'. A particular value of the random variable will be denoted by using a subscript, e.g., p_0'. A capital letter is not used since P has been used to denote a probability.

$$p' = \frac{64}{100} = 0.64$$

Hence $$\sigma_{p'} = \sqrt{\frac{(0.64)(0.36)}{100}} = 0.048$$

A 99% confidence interval for p would be as follows:

$$0.64 - (2.58)(0.048) < p < 0.64 + (2.58)(0.048)$$

or $$0.52 < p < 0.76$$

It will be noted that the interval stated above is a rather wide one, but this is partly a consequence of the high level of confidence. If a lower level of confidence were acceptable, then a shorter interval would be obtained.

7.6 Tests of Hypotheses Concerning a Population Mean

The nature of the testing of hypotheses has been outlined in Section 7.1. As was indicated there, a hypothesis is an assertion concerning a population parameter. Furthermore, this assertion is accompanied by a contrary assertion called the alternative hypothesis. The hypothesis is usually indicated by the symbol H_0 and the alternative by H_1. For example, a typical hypothesis and its alternative might be as follows:

$$H_0 : \mu = 40$$
$$H_1 : \mu < 40$$

As has been indicated, a test of a hypothesis is a procedure for arriving at a decision to either accept it as true or to reject it as false. Rejecting the hypothesis is, of course, tantamount to accepting the alternative.

It is clear that in reaching a decision of this kind, two kinds of error are possible:

(1) Rejecting a true hypothesis

(2) Accepting a false hypothesis

The probability of making a type 1 error is usually designated by the letter α, whereas the probability of making a type 2 error is designated by the letter β.

The procedure for testing the previously stated hypothesis concerning the value of the mean will involve the drawing of a random sample and the calculation of the sample mean. A sample mean near 40 will tend to support the hypothesis and lead to the acceptance of the hypothesis. On the other hand, a sample mean much less than 40 will cast doubt on the hypothesis and lead to its rejection in favor of the alternative.

The question that naturally arises is just how far below 40 the sample mean can be before the hypothesis is rejected in favor of the alternative. In other words, what is the critical value $\bar{x}_0$ such that if $\bar{X} > \bar{x}_0$, H_0 will be accepted, but if $\bar{X} < \bar{x}_0$, H_0 will be rejected?

The answer to this question depends upon the size of the error probabilities α and β one is willing to tolerate. Let us suppose, for example, that the value of α is to be set at 0.05; i.e., the test is to be conducted in such a way that there will be a 0.05 probability that H_0 will be rejected when $\mu = 40$. Suppose also that the sample size is to be 36 and that the standard deviation of the population is known to be approximately 9.* Then $\bar{x}_0$ is to be determined such that

$$P(\bar{X} < \bar{x}_0 | \mu = 40) = 0.05$$

But

$$P(\bar{X} < \bar{x}_0 | \mu = 40) = P(Z < z_0)$$

where

$$z_0 = \frac{\bar{x}_0 - \mu}{\sigma/\sigma_n} = \frac{\bar{x}_0 - 40}{9/\sqrt{36}} = \frac{\bar{x}_0 - 40}{1.5}$$

The table of normal curve areas shows that if

$$P(Z < z_0) = 0.05$$

then

$$z_0 \simeq -1.64$$

Therefore

$$\frac{\bar{x}_0 - 40}{1.5} = -1.64$$

or

$$\begin{aligned} \bar{x}_0 &= (1.5)(-1.64) + 40 \\ &= -2.46 + 40 \\ &= 37.54 \end{aligned}$$

Hence our test procedure has been established as follows:

If $\bar{x} < 37.54$, reject H_0.

If $\bar{x} \geq 37.54$, accept H_0.

* This knowledge might be based on a previous sample calculation.

Although this test satisfies the specified requirement that the probability of rejecting H_0 when it is true is 0.05, it has not yet been assessed from the standpoint of the probability of the type 2 error, i.e., the probability of accepting the hypothesis $\mu = 40$ when $\mu < 40$. We shall now analyze the test from that standpoint.

It is clear at the outset that the statement that H_0 is false does not imply that μ has any specific value, but rather that it has some value less than 40. Hence, the value of β is not unique. In fact, since β is a function of μ, it is appropriate to designate it as $\beta(\mu)$. We shall now proceed to calculate several values of $\beta(\mu)$.

$$\beta(38) = P(\bar{X} \geq 37.4 | \mu = 38) = P\left(Z \geq \frac{37.4 - 38}{1.5}\right)$$
$$= P(Z \geq -0.4) = 0.66$$

$$\beta(36) = P(\bar{X} \geq 37.4 | \mu = 36) = P\left(Z \geq \frac{37.4 - 36}{1.5}\right)$$
$$= P(Z \geq 0.93) = 0.18$$

$$\beta(34) = P(\bar{X} \geq 37.4 | \mu = 34) = P\left(Z \geq \frac{37.4 - 34}{1.5}\right)$$
$$= P(Z \geq 2.3) = 0.01$$

It will be observed that, as would be expected, the value of β drops sharply as the assumed alternative value of μ decreases. In other words, the farther the true value of μ is below the value stated in H_0, the less likely it is that the error of accepting H_0 will occur. The rather high value of $\beta(38)$ may not necessarily indicate that the test is unsatisfactory, since accepting the hypothesis that $\mu = 40$ when, in fact, $\mu = 38$ may not be a serious matter. If, in fact, the situation is such that the acceptance of the hypothesis does not constitute a serious error as long as $\mu \geq 36$, then since $\beta(36) = 0.18$, it may be stated that there is less than a 1 in 5 chance of making a serious type 2 error.

The test that has just been used as an illustration is referred to as a **one-tail test** in view of the fact that the rejection region was confined to the lower end, or *tail,* of the normal curve. We shall now illustrate a **two-tail test,** i.e., one in which the rejection region is divided between both tails of the curve. Let

$$H_0 : \mu = 20$$
$$H_1 : \mu \neq 20$$

We shall suppose that this test is to be conducted using a sample of size 64 and that $\sigma = 4$.

In this case a value of $\bar{x}$ either much greater than or much less than 20 will cast doubt on the truth of the hypothesis. If it is specified that $\alpha = 0.05$, then we wish to determine $\bar{x}_1$ and $\bar{x}_2$ such that

$$P(\bar{X} < \bar{x}_1 \text{ or } \bar{X} > \bar{x}_2 | \mu = 20) = 0.05$$

But $$P(\bar{X} < \bar{x}_1 \text{ or } \bar{X} > \bar{x}_2 | \mu = 20) = P(Z < z_1 \text{ or } Z > z_2)$$

where $$z_1 = \frac{\bar{x}_1 - 20}{4/\sqrt{64}} = \frac{\bar{x}_1 - 20}{0.5}$$

and $$z_2 = \frac{\bar{x}_2 - 20}{0.5}$$

But the table of normal curve areas shows that the values $z_1 = -2.0$, $z_2 = 2.0$ will approximately satisfy the equation

$$p(Z < z_1 \text{ or } Z > z_2) = 0.05$$

Therefore $$\frac{\bar{x}_1 - 20}{0.5} = -2.0 \qquad \frac{\bar{x}_2 - 20}{0.5} = 2.0$$

or $$\bar{x}_1 = 19.0, \qquad \bar{x}_2 = 21.0$$

Hence our test procedure has been established as follows:

If $\bar{x} < 19.0$ or $\bar{x} > 21.0$, reject H_0.

If $19.0 \leq \bar{x} \leq 21.0$, accept H_0.

To illustrate the computation of β for the two-tail test, we shall compute $\beta(22)$.

$$\begin{aligned}\beta(22) &= P(19.0 \leq \bar{X} \leq 21.0 | \mu = 22)\\ &= P\left(\frac{19.0 - 22}{0.5} < Z < \frac{21.0 - 22}{0.5}\right)\\ &= P(-6 < Z -2.0)\\ &\simeq 0.02\end{aligned}$$

7.7 Tests of Hypotheses Concerning a Population Proportion

The principles involved in testing hypotheses concerning the value of a population proportion p are similar to those employed in testing hypotheses about the mean. Consider the following example.

Let

$$H_0: p = 0.64$$
$$H_1: p > 0.64$$

Suppose the hypotheses is to be tested using a sample of size 100, with $\alpha = 0.01$. We wish to choose p_0' such that

$$P(p' > p_0' | p = 0.64) = 0.01$$

But

$$P(p' > p_0' | p = 0.64) = P(Z > z_0)$$

where

$$z_0 = \frac{p_0' - p}{\sqrt{\dfrac{pq}{n}}} = \frac{p_0' - 0.64}{\sqrt{\dfrac{(0.64)(0.36)}{100}}} = \frac{p_0' - 0.64}{0.048}$$

But the table of the normal curve areas shows that if

$$P(Z > z_0) = 0.01$$

then

$$z_0 = 2.33$$

Therefore

$$\frac{p_0' - 0.64}{0.048} = 2.33$$

or

$$p_0' = (2.33)(0.048) + 0.64 = 0.75$$

Hence our test procedure is as follows:

If $p' > 0.75$, reject H_0.

If $p' \leq 0.75$, accept H_0.

Two-tail tests of hypotheses concerning a population proportion are conducted in a manner similar to those involving the mean. Two values of p', p_1', and p_2' are obtained such that if $\alpha = \alpha_0$, then

$$P(p' < p_1' \text{ or } p' > p_2' | H_0 \text{ is true}) = \alpha_0$$

and the test procedure is as follows:

If $p' < p_1'$ or $p' > p_2'$, reject H_0.

If $p_1' \leq p' \leq p_2'$, accept H_0.

An example of this kind will be found in the exercises.

7.8 Small Sample Tests

All of the tests discussed thus far have been **large sample tests,** involving samples of at least 30. This has enabled us to assume that either the sample mean or the sample proportion was at least approximately normally distributed. Furthermore, it enables us to use the sample standard deviation as an estimate of the corresponding population value.

For cases in which samples are smaller than 30, **small sample methods** must be used in obtaining confidence intervals or for testing hypotheses concerning a population mean. These methods involve the distribution of the random variable T, where

$$T = \frac{\overline{X} - \mu}{s/\sqrt{n}}$$

It will be noted that the calculation of T involves the sample standard deviation s rather than the population standard deviation σ. Theoretically, the use of T requires the assumption that the population is normally distributed, but in practice moderate departures from normality will not seriously affect the results. Tables of the distribution of T will be found in most books on statistical inference.

Problems—Chapter 7

1. Given the sample values 5, 1, 2, 3, 9, find:
   (a) the mean
   (b) the median
   (c) the mean deviation
   (d) the variance
   (e) the standard deviation
2. A sample of size 36 is to be drawn from a population with mean 20 and standard deviation 4. Find the probability that the sample mean will exceed 21.
3. A sample of size 64 is drawn from a population whose standard deviation is 4, and the sample mean is found to be 30. Determine a 99% confidence interval for the population mean.
4. Using the data of Problem 3, determine a 90% confidence interval for the population mean.
5. A random sample of 100 manufactured articles is drawn, and it is observed that 20 are defective. Find a 95% confidence interval for the proportion of defectives in the population.
6. For a population whose standard deviation is 5, a test of the following hypothesis is to be made, using a sample of size 100, with $\alpha = 0.05$:

   $$H_0: \mu = 30$$
   $$H_1: \mu > 30$$

   For what values of $\overline{X}$ will H_0 be rejected?

7. Answer the question posed in Problem 6 if H_1 is changed to $\mu \neq 30$.
8. Using the data and hypotheses of Problem 6, calculate β (32) and state the meaning of the value that you have obtained.
9. A test of the following hypothesis is to be made, using a sample of size 100, with $\alpha = 0.05$:

$$H_0: p = 0.80$$
$$H_1: p < 0.80$$

For what values of the sample proportion p' will H_0 be rejected?
10. Answer the question posed in Problem 9 if H_1 is changed to $p \neq 0.80$.

8

BASIC LINEAR ALGEBRA

8.1 Three-Dimensional Space

In Section 1.2 we established a correspondence between the real numbers and the points on a straight line. In Section 1.3 we introduced the rectangular coordinate system, which, using two perpendicular lines, established a correspondence between the set of ordered pairs of real numbers and the points in a plane. By adding a third line, perpendicular to each of the other two, we obtain a three-dimensional rectangular coordinate system in which there is a one-to-one correspondence between the set of ordered triples of real numbers and the points in space.

If we think of the x and y axes as lying in a horizontal plane, then the third axis, which we shall designate as the z axis, will extend in the vertical direction above and below the xy plane and will pass through the point of intersection of the x and y axes. This point will be called the origin of the three-dimensional coordinate system. The positive z axis will extend upward from the origin, the negative z axis downward.

A point whose coordinates are specified by an ordered triple may be located as follows. First locate the point in the xy plane that corresponds to the first two coordinates. Then proceed in the vertical direction to a point opposite the point on the z axis specified by the third coordinate of the point.

Figure 8.1 shows a three-dimensional rectangular coordinate system in which the arrows indicate the positive direction on each axis. A plotted point is also shown.

It will be recalled that in connection with the two-dimensional rectangular coordinate system we made use of an important formula called the distance formula, which enabled us to calculate the distance between any two specified points. This formula may be readily expanded to the three-dimensional case. If the coordinates of two

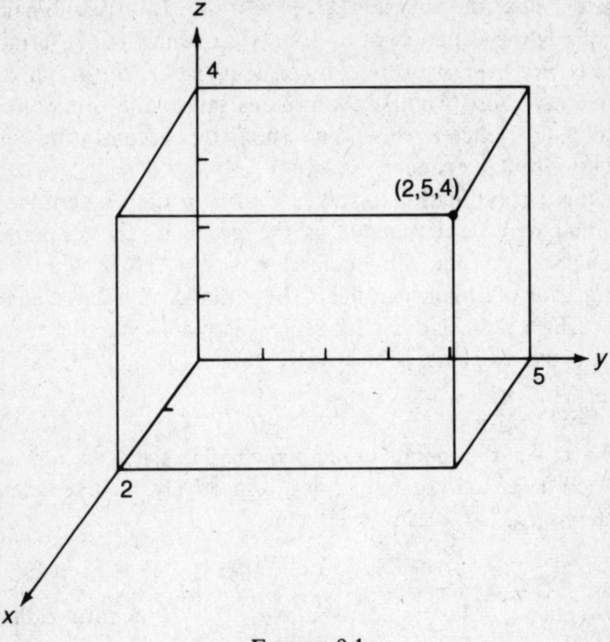

FIGURE 8.1

points are (x_1,y_1,z_1) and (x_2,y_2,z_2), then the distance between them is given by

$$d = \sqrt{(x_2 - x_1)^2 + (y_2 - y_1)^2 + (z_2 - z_1)^2}$$

For example, the distance between the points (2,4,−3) and (6,−7,4) is

$$\sqrt{(6-2)^2 + (-7-4)^2 + (4+3)^2} = \sqrt{186} \approx 13.6$$

Furthermore, since a **sphere** is defined as the set of all points in space whose distance from a fixed point called the center is a constant called the radius, the equation of a sphere whose center is (x_1,y_1,z_1) and whose radius is r may be written:

$$\sqrt{(x - x_1)^2 + (y - y_1)^2 + (z - z_1)^2} = r$$

or

$$(x - x_1)^2 + (y - y_1)^2 + (z - z_1)^2 = r^2$$

In the special case in which the center is the origin this becomes

$$x^2 + y^2 = z^2 = r^2$$

Consider next the equation $x^2 + y^2 = 4$. In a two-dimensional space this is the equation of a circle with center at the origin and radius 2. However, in a three-dimensional space, the graph consists of all points whose x and y coordinates satisfy the equation with z unrestricted. Hence in three dimensions this becomes the equation of a **right circular cylinder.**

The three coordinate axes taken two at a time determine three planes that may be designated as the xy plane, the xz plane, and the yz plane. An equation such as $x = k$, where k is a constant, is the equation of a plane parallel to the yz plane. Similarly, equations of the form $y = k$ and $z = k$ represent planes parallel to the xz plane and the xy plane, respectively.

8.2 Vectors

We have seen that points in the plane and in space are represented by ordered pairs and ordered triples, respectively. These are special cases of ordered n-tuples of the form

$$(a_1,a_2, \ldots, a_n)$$

called **vectors.** The numbers $a_1, a_2, \ldots, a_n$ are called the **components** of the vector.

The components of a vector do not necessarily represent geometric coordinates. For example, if the unit prices of four commodities are \$4, \$6, \$3, and \$7, respectively, this fact might be represented by the vector (4,6,3,7). The vector (37,42,36) might be associated with the numbers of people at a party at three given times. The operations of addition and multiplication are defined in a special way for vectors, which are denoted by boldface letters. Two vectors are said to be of the same type if they have the same number of components.

Given two vectors $\mathbf{A} = (a_1,a_2, \ldots, a_n)$ and $\mathbf{B} = (b_1,b_2, \ldots, b_n)$, the sum of $\mathbf{A}$ and $\mathbf{B}$ is defined as follows:

$$\mathbf{A} + \mathbf{B} = (a_1 + b_1,\ a_2 + b_2, \ldots, a_n + b_n)$$

Hence the sum of two vectors of the same type is a vector whose components are the sums of the corresponding components in the vectors being added.

The negative of a vector $\mathbf{A} = (a_1,a_2, \ldots, a_n)$ is the vector $-\mathbf{A} = (-a_1,-a_2, \ldots, -a_n)$. The difference of two vectors $\mathbf{A} - \mathbf{B}$ is then defined as the sum of the vectors $\mathbf{A}$ and $-\mathbf{B}$. Hence if

$\mathbf{A} = (a_1,a_2, \ldots, a_n)$ and $\mathbf{B} = (b_1, b_2, \ldots, b_n)$, then $\mathbf{A} - \mathbf{B} = (a_1 - b_1, a_2 - b_2, \ldots, a_n - b_n)$.

In defining products we note first that a real number is often called a **scalar** to distinguish it from a vector. If $\mathbf{A} = (a_1,a_2, \ldots, a_n)$ and c is a scalar, then the product of c and $\mathbf{A}$ is defined as

$$c\mathbf{A} = (ca_1, ca_2, \ldots, ca_n)$$

Hence, the product of a scalar and a vector is a vector whose components are obtained by multiplying each component of the given vector by the scalar.

Next we define a product of two vectors, which is sometimes called the **scalar product** since the multiplication of the two vectors produces a scalar in this case. Given two vectors $\mathbf{A} = (a_1,a_2, \ldots, a_n)$ and $\mathbf{B} = (b_1,b_2, \ldots, b_n)$, the scalar product is designated by $\mathbf{A} \cdot \mathbf{B}$ and is defined as follows:

$$\mathbf{A} \cdot \mathbf{B} = a_1b_1 + a_2b_2 + \ldots + a_nb_n$$

Hence the scalar product of two vectors of the same type is a scalar equal to the sum of the products of the corresponding components of the two vectors. The scalar product is sometimes referred to as the **dot product.**

There is a second type of product of two vectors, called the **vector product,** but it is defined only for vectors having three components, and we will not discuss it in this book.

Another useful vector concept is that of the **norm** of a vector $\mathbf{A}$ denoted by $\|\mathbf{A}\|$. By definition

$$\|\mathbf{A}\| = \sqrt{a_1^2 + a_2^2 + \ldots + a_n^2}$$

Hence the norm of a vector is a scalar equal to the square root of the sum of the squares of the components of a vector.

Example 1 Given $\mathbf{A} = (3,-2,4,5)$ and $\mathbf{B} = (2,7,-1,6)$. Find

(a) $\mathbf{A} + \mathbf{B}$ (c) $\mathbf{A} \cdot \mathbf{B}$

(b) $4\mathbf{A}$ (d) $\|\mathbf{A}\|$

Solution

(*a*) $\mathbf{A} + \mathbf{B} = (3 + 2, -2 + 7, 4 - 1, 5 + 6) = (5,5,3,11)$

(*b*) $4\mathbf{A} = (12,-8,16,20)$

(*c*) $\mathbf{A} \cdot \mathbf{B} = (3)(2) + (-2)(7) + (4)(-1) + (5)(6) = 18$

(*d*) $\|\mathbf{A}\| = \sqrt{9 + 4 + 16 + 25} = \sqrt{54} = 3\sqrt{6}$

The equality of two vectors is defined as follows:

If $\mathbf{A} = (a_1,a_2, \ldots, a_n)$ and $\mathbf{B} = (b_1,b_2, \ldots, b_n)$, then $\mathbf{A} = \mathbf{B}$ if and only if $a_1 = b_1$, $a_2 = b_2$, . . . , $a_n = b_n$. Hence two vectors are equal if and only if their respective components are equal. The vector whose components are all zeros is called the **zero vector.**

The following laws of vector algebra can be easily verified.

(1) $\mathbf{A} + \mathbf{B} = \mathbf{B} + \mathbf{A}$

(2) $\mathbf{A} \cdot \mathbf{B} = \mathbf{B} \cdot \mathbf{A}$

(3) $\mathbf{A} \cdot (\mathbf{B} + \mathbf{C}) = \mathbf{A} \cdot \mathbf{B} + \mathbf{A} \cdot \mathbf{C}$

(4) $\mathbf{A} \cdot \mathbf{A} = (\|\mathbf{A}\|)^2$

With somewhat more difficulty the following inequality can be established:

(5) $\|\mathbf{A} + \mathbf{B}\| \leq \|\mathbf{A}\| + \|\mathbf{B}\|$

8.3 Applications of Vectors

One of the most common interpretations of vectors has already been alluded to in Section 8.1. It involves the geometric concept of a vector. For example, the vector (2,3,4) may be interpreted as a point in three-dimensional space. However, it may also be interpreted as a directed line segment, or arrow, extending from the origin to the point (2,3,4) as shown in Figure 8.2.

A third possible interpretation is that of an arrow extending *from* a point (x_1,y_1,z_1) *to* a point $(x_1 + 2, y_1 + 3, z_1 + 4)$.

The geometric interpretation of vectors as arrows in space is an extremely useful one and has important application to physical problems.

To illustrate this application of vectors, it is first necessary to show the geometric interpretation of a vector sum. If **A** and **B** are two vectors represented by arrows originating at the same point, then it is readily seen that the vector $\mathbf{A} + \mathbf{B}$ is the diagonal of the parallelogram determined by **A** and **B.** This fact is illustrated in Figure 8.3.

To see how this fact may be used in physical problems, it should be noted that many physical quantities are vector quantities in the sense that their specification requires both a magnitude and a direction.

An example of this type of application arises in a situation in which two forces, each with a specified magnitude and direction,

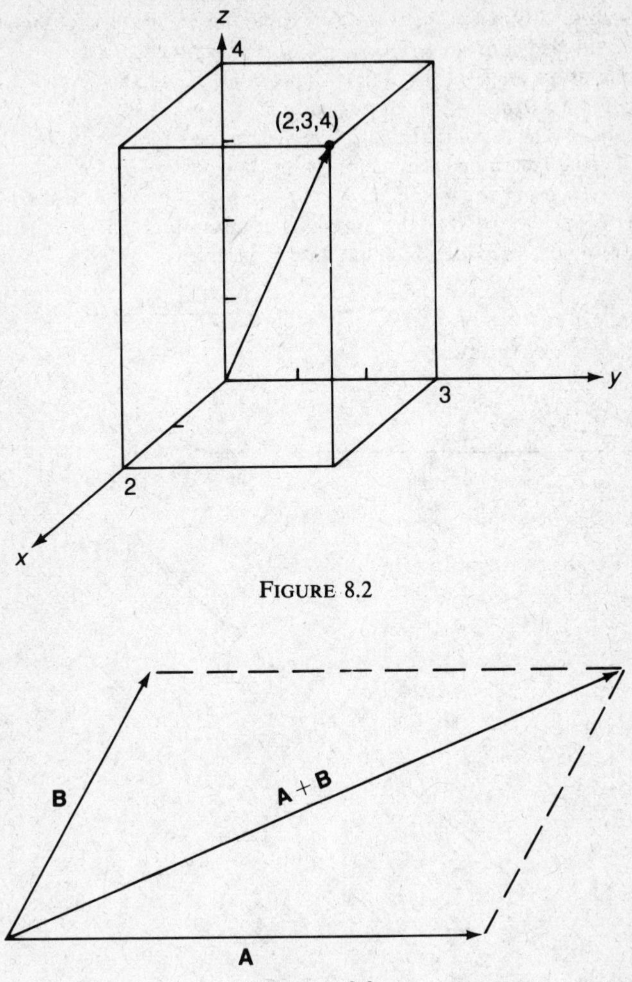

FIGURE 8.2

FIGURE 8.3

act simultaneously on an object and it is desired to find the **resultant force,** i.e., the single force that produces the combined effect of the two given forces. Each of the given forces may be represented by a geometric vector whose length is proportional to the magnitude of the force and that points in the specified direction. The resultant force will then be indicated by the vector representing the sum of the given force vectors, i.e., the diagonal of the parallelogram determined by these two vectors.

Example 1 Suppose that one force acting on an object is of magnitude 15 lb in the horizontal direction and that another is a force of 10 lb acting at an angle of 30° to the horizontal. Find the resultant force. The force diagram is shown in Figure 8.4.

Solution The resultant force can be determined approximately by measuring the length and inclination of the diagonal vector or, more accurately, by using trigonometry. Using the law of cosines (Section 3.10), it is readily found that the approximate magnitude and inclination of the resultant force are 22 lb and 23°, respectively.

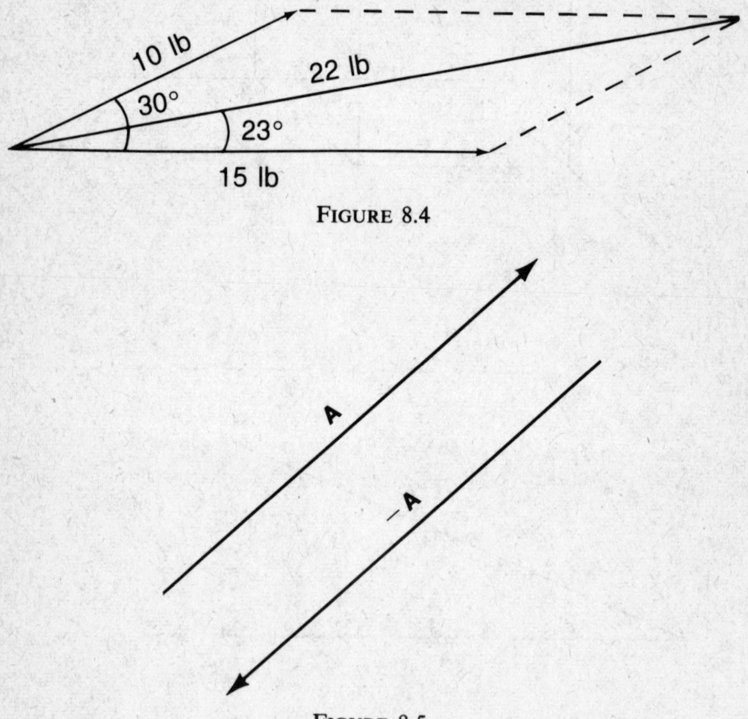

FIGURE 8.4

FIGURE 8.5

If a given vector is represented by an arrow, then the negative of the given vector is represented by an arrow pointing in the opposite direction. See Figure 8.5.

The difference between two vectors **A** − **B** may be represented geometrically as the sum of the vectors **A** and −**B**, as shown in Figure 8.6.

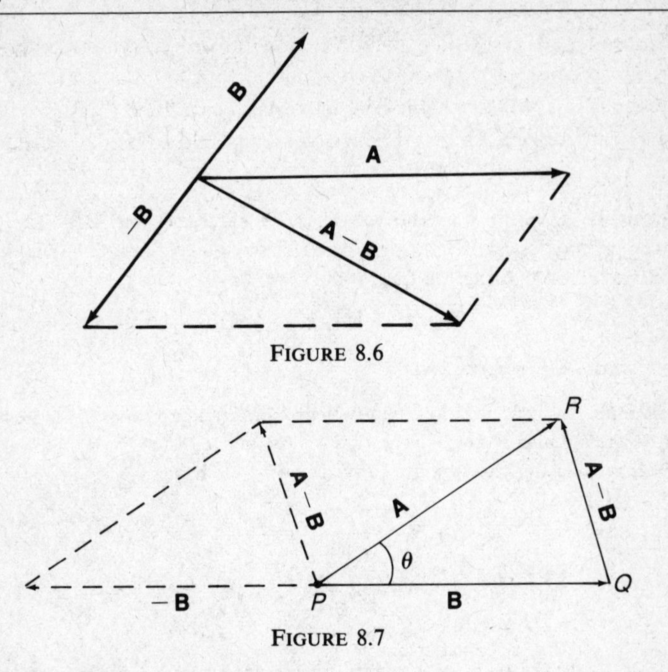

FIGURE 8.6

FIGURE 8.7

It is readily seen that for a given vector $\mathbf{A} = (a_1,a_2,a_3)$, $\|\mathbf{A}\|$ is the length of the arrow representing the vector $\mathbf{A}$ since $\sqrt{a_1^2 + a_2^2 + a_3^2}$ is the distance from the origin to the point (a_1,a_2,a_3).

The dot product and the norms of two given vectors may be used to determine the angle θ between the vectors, where $0 \leq \theta \leq 180°$.

In Figure 8.7 the vectors $\mathbf{A}$ and $\mathbf{B}$ are shown originating at the vertex P of the triangle PQR. Using the law of cosines, we have

$$(\|\mathbf{A} - \mathbf{B}\|)^2 = (\|\mathbf{A}\|)^2 + (\|\mathbf{B}\|)^2 - 2\,\|\mathbf{A}\|\,\|\mathbf{B}\| \cos\theta$$

or

$$\cos\theta = \frac{(\|\mathbf{A}\|)^2 + (\|\mathbf{B}\|)^2 - (\|\mathbf{A} - \mathbf{B}\|)^2}{2\,\|\mathbf{A}\|\,\|\mathbf{B}\|}$$

$$= \frac{\mathbf{A}\cdot\mathbf{A} + \mathbf{B}\cdot\mathbf{B} - (\mathbf{A} - \mathbf{B})\cdot(\mathbf{A} - \mathbf{B})}{2\|\mathbf{A}\|\,\|\mathbf{B}\|}$$

$$= \frac{\mathbf{A}\cdot\mathbf{A} + \mathbf{B}\cdot\mathbf{B} - \mathbf{A}\cdot\mathbf{A} + \mathbf{A}\cdot\mathbf{B} + \mathbf{A}\cdot\mathbf{B} - \mathbf{B}\cdot\mathbf{B}}{2\|\mathbf{A}\|\,\|\mathbf{B}\|}$$

$$= \frac{2\mathbf{A}\cdot\mathbf{B}}{2\|\mathbf{A}\|\,\|\mathbf{B}\|} = \frac{\mathbf{A}\cdot\mathbf{B}}{\|\mathbf{A}\|\,\|\mathbf{B}\|}$$

Since $0 \leq \theta \leq 180°$, $\theta = 90°$ if and only if $\cos \theta = 0$. Hence two vectors are perpendicular to each other if and only if their dot product is 0. Note that this condition will be fulfilled if either vector is the zero vector. Hence the zero vector is said to be perpendicular to every vector.

Example 2 Show that the vectors $\mathbf{A} = (12,2,-4)$ and $\mathbf{B} = (2,-2,5)$ are perpendicular to each other.

Solution Computing the dot product, we have

$$\mathbf{A} \cdot \mathbf{B} = 24 - 4 - 20 = 0$$

which proves the assertion.

Example 3 Find, to the nearest degree, the angle between the vectors $\mathbf{A} = (1,2,4)$ and $\mathbf{B} = (-2,3,1)$.

Solution Using the formula proved above, we have

$$\cos \theta = \frac{\mathbf{A} \cdot \mathbf{B}}{\|\mathbf{A}\| \; \|\mathbf{B}\|}$$

$$= \frac{-2+6+4}{\sqrt{21}\,\sqrt{14}} \simeq 0.47$$

Using a table of the trigonometric functions, we find $\theta \simeq 62°$.

We now note two nongeometric applications of vectors, or more specifically, of the dot product.

Suppose the vector $\mathbf{C} = (c_1,c_2,c_3,c_4)$, is a cost vector whose components represent the unit cost of each of four commodities and that the vector $\mathbf{N} = (n_1,n_2,n_3,n_4)$ is a purchase vector whose components represent the number of units of each of the commodities purchased. Then $\mathbf{N} \cdot \mathbf{C} = n_1c_1 + n_2c_2 + n_3c_3 + n_4c_4$ has the evident interpretation of the total cost of the combined purchase.

Another useful application of the dot product occurs in the field of probability. Suppose the five possible values of a random variable are represented by the vector $\mathbf{X} = (x_1,x_2,x_3,x_4,x_5)$ and that their respective probabilities are represented by the vector $\mathbf{P} = (p_1,p_2,p_3,p_4,p_5)$. Then, recalling the definition of the expected value $E(X)$, (Section 4.10), we readily see that $E(X) = \mathbf{P} \cdot \mathbf{X}$.

8.4 Matrices

We have seen that a vector is an ordered n-tuple and that such quantities are useful in representing data such as the unit costs of several commodities when the position of a cost in the n-tuple identifies it with a specified commodity. This idea may be readily extended to a two-way array such as the following:

	I	II	III	IV
Jan.	20	15	32	41
Feb.	22	17	30	42
Mar.	21	19	35	39

Such an array is called a **matrix.** The horizontal sets are called **rows** and the vertical sets are called **columns.** Here the columns might be identified with four commodities and the rows with the first days of the months January, February, and March. Hence the element 30 of the matrix that appears in the second row and third column is the cost of the third commodity on February 1.

In general, a matrix is a rectangular array of numbers, called the **elements** of the matrix. The element in the ith row and the jth column is designated by a_{ij}: the first subscript refers to the column and the second to the row. A matrix with m rows and n columns is called an $m \times n$ matrix (read m by n matrix) and may be exhibited as follows:

$$\begin{bmatrix} a_{11} & a_{12} & \cdots & a_{1n} \\ a_{21} & a_{22} & \cdots & a_{2n} \\ a_{m1} & a_{m2} & \cdots & a_{mn} \end{bmatrix}$$

Evidently a vector may be regarded as a special case of a matrix. A matrix consisting of only one row is called a **row vector,** a matrix consisting of only one column is called a **column vector.**

Two matrices are said to be of the same type if they contain the same number of rows and columns, respectively.

A matrix with the same number of rows as columns is called a **square matrix.** A square ($n \times n$) matrix in which $a_{ij} = a_{ji}$, $i = 1, 2, \ldots n$, $j = 1, 2, \ldots n$, is called a **symmetric matrix.** For example, the matrix

$$\begin{bmatrix} 2 & 3 & 5 \\ 3 & 1 & 6 \\ 5 & 6 & 7 \end{bmatrix}$$

is a symmetric matrix. A matrix in which each element is 0 is called the **zero matrix.**

The **transpose** of a given matrix A, denoted by A^t, is the matrix obtained by interchanging the rows and columns of a matrix. Hence if

$$A = \begin{bmatrix} a_{11} & a_{12} \\ a_{21} & a_{22} \end{bmatrix}$$

then
$$A^t = \begin{bmatrix} a_{11} & a_{21} \\ a_{12} & a_{22} \end{bmatrix}$$

It is easily seen that for a symmetric matrix, $A = A^t$.

The elements $a_{11}, a_{22}, \ldots, a_{nn}$ of a square matrix comprise the **principal diagonal** of the matrix. A square matrix in which the elements of the principal diagonal are 1's and in which all of the other elements are 0's is called an **identity matrix.** The $n \times n$ identity matrix is denoted by I_n, for example,

$$I_3 = \begin{bmatrix} 1 & 0 & 0 \\ 0 & 1 & 0 \\ 0 & 0 & 1 \end{bmatrix}$$

If the size of the matrix is clear from the context, then the identity matrix is sometimes designated simply by I.

We shall now define some operations of matrix algebra, beginning with the operation of addition, which is defined only for matrices of the same type.

If A and B are two matrices of the same type, then the sum $A + B$ is a matrix in which each element is the sum of the corresponding elements in A and B. Hence for two 2×3 matrices we have

$$\begin{bmatrix} a_{11} & a_{12} & a_{13} \\ a_{21} & a_{22} & a_{23} \end{bmatrix} + \begin{bmatrix} b_{11} & b_{12} & b_{13} \\ b_{21} & b_{22} & b_{23} \end{bmatrix}$$
$$= \begin{bmatrix} a_{11} + b_{11} & a_{12} + b_{12} & a_{13} + b_{13} \\ a_{21} + b_{21} & a_{22} + b_{22} & a_{23} + b_{23} \end{bmatrix}$$

Next we define multiplication of a matrix by a scalar. If c is a ar and A is a matrix, then the product cA is a matrix in which element is the product of c and the corresponding element in hus:

$$c\begin{bmatrix} a_{11} & a_{12} & a_{13} \\ a_{21} & a_{22} & a_{23} \end{bmatrix} = \begin{bmatrix} ca_{11} & ca_{12} & ca_{13} \\ ca_{21} & ca_{22} & ca_{23} \end{bmatrix}$$

that each of the operations defined thus far has the commuta-erty, i.e., $A + B = B + A$ and $cA = Ac$.

We now define an operation that is not commutative, the multiplication of two matrices. This operation is defined only in the case in which the number of columns in the first factor of the product equals the number of rows in the second factor.

If A is an $m \times n$ matrix and B is an $n \times q$ matrix, then the product AB is an $m \times q$ matrix in which the element in the ith row and the jth column is the sum of the products obtained by multiplying each of the elements in the ith row of A by the corresponding element in the jth column of B. Hence, if the element in the ith row and the jth column of the product matrix is designated by c_{ij}, then

$$c_{ij} = a_{i1}b_{1j} + a_{i2}b_{2j} + \cdots + a_{in}b_{nj}$$

The following example will serve to illustrate this definition:

$$\begin{bmatrix} 2 & 3 \\ 6 & 4 \\ 7 & 6 \end{bmatrix} \begin{bmatrix} 5 & 1 & 3 \\ 4 & 2 & 8 \end{bmatrix} = \begin{bmatrix} 22 & 8 & 30 \\ 46 & 14 & 50 \\ 59 & 19 & 69 \end{bmatrix}$$

Note, for example, that the element in the first row and the first column of the product is calculated as follows:

$$(2)(5) + (3)(4) = 22$$

whereas the element in the third row and the second column is obtained by

$$(7)(1) + (6)(2) = 19$$

The student may readily verify the calculation of the other elements of the product matrix.

As previously stated, matrix multiplication is not commutative, i.e., in general, $AB \neq BA$. This fact may be easily verified using the above illustration. If the order in which the matrices is multiplied is reversed, we have

$$\begin{bmatrix} 5 & 1 & 3 \\ 4 & 2 & 8 \end{bmatrix} \begin{bmatrix} 2 & 3 \\ 6 & 4 \\ 7 & 6 \end{bmatrix} = \begin{bmatrix} 38 & 37 \\ 76 & 68 \end{bmatrix}$$

Note that a row vector may be multiplied by a column vector provided that they have the same number of elements, for example,

$$[1\ 6\ 5\ 3]\begin{bmatrix}5\\4\\2\\1\end{bmatrix}=[42]$$

Note that the single element of the product is the dot product of the two vectors. However, it should be emphasized that the dot product is a scalar, whereas the matrix product is a matrix consisting of a single element. The distinction becomes clear when it is observed that a scalar may be multiplied by any matrix, whereas a matrix consisting of a single element may be the first factor in a product with only a one-row matrix, i.e., a row vector. Note also that for any square matrix A, $AI = IA = A$. This fact accounts for the designation of the matrix I as the identity matrix. For example, it is readily seen that

$$\begin{bmatrix}1&0&0\\0&1&0\\0&0&1\end{bmatrix}\begin{bmatrix}2&4&1\\1&6&5\\3&7&8\end{bmatrix}=\begin{bmatrix}2&4&1\\1&6&5\\3&7&8\end{bmatrix}$$

In fact, the identity property holds for the product of any matrix and an identity matrix provided that the requirements of matrix multiplication are met.

Given a square matrix A, if there exists a matrix A^{-1} such that $AA^{-1} = A^{-1}A = I$, then A^{-1} is said to be the **inverse** of the matrix A. For a given matrix, an inverse does not necessarily exist. However, if an inverse does exist, it is unique.

Suppose, for example, we wish to find the inverse of the matrix

$$A=\begin{bmatrix}2&4\\1&3\end{bmatrix}$$

If the inverse is designated by

$$A^{-1}=\begin{bmatrix}a_{11}&a_{12}\\a_{21}&a_{22}\end{bmatrix}$$

then we wish to determine the elements of A^{-1} such that

$$\begin{bmatrix}2&4\\1&3\end{bmatrix}\begin{bmatrix}a_{11}&a_{12}\\a_{21}&a_{22}\end{bmatrix}=\begin{bmatrix}1&0\\0&1\end{bmatrix}$$

Using the definition of matrix multiplication, it is clear that the above matrix equation is equivalent to the following system of linear equations.

$$2a_{11} + 4a_{21} = 1$$
$$2a_{12} + 4a_{22} = 0$$
$$a_{11} + 3a_{21} = 0$$
$$a_{12} + 3a_{22} = 1$$

This system can be readily solved by the usual methods. Note that the first and third equations involve only a_{11} and a_{21}, whereas the second and fourth involve only a_{12} and a_{22}. We obtain

$$a_{11} = \tfrac{3}{2} \qquad a_{12} = -2 \qquad a_{21} = -\tfrac{1}{2} \qquad a_{22} = 1$$

Hence
$$A^{-1} = \begin{bmatrix} \frac{3}{2} & -2 \\ -\frac{1}{2} & 1 \end{bmatrix}$$

That the inverse of a matrix does not necessarily exist may be demonstrated by the following example:

Let
$$A = \begin{bmatrix} 2 & 4 \\ 1 & 2 \end{bmatrix}$$

If the inverse of A, if it exists, is designated by the matrix

$$\begin{bmatrix} a_{11} & a_{12} \\ a_{21} & a_{22} \end{bmatrix}$$

then we would determine A^{-1} by solving the system of equations:

$$2a_{11} + 4a_{21} = 1$$
$$2a_{12} + 4a_{22} = 0$$
$$a_{11} + 2a_{21} = 0$$
$$a_{12} + 2a_{22} = 1$$

It is readily seen that the first and third equations are inconsistent, as are the second and fourth. Hence the system has no solution, and the inverse of A does not exist.

It should be noted that although the procedure shown above will always produce the inverse of a matrix if it exists, more efficient methods are available for its computation.

8.5 Determinants

A **determinant** is a function of a square matrix. The general definition of a determinant is rather complex. However, we shall initially consider only the definitions of the determinants of 2×2 and 3×3 matrices.

The determinant of the matrix

$$\begin{bmatrix} a_{11} & a_{12} \\ a_{21} & a_{22} \end{bmatrix}$$

is defined to be the number $a_{11}a_{22} - a_{12}a_{21}$. The determinant of the above matrix is often designated by the symbol

$$\begin{vmatrix} a_{11} & a_{12} \\ a_{21} & a_{22} \end{vmatrix}$$

Hence for the matrix

$$\begin{bmatrix} 2 & 3 \\ -4 & 7 \end{bmatrix}$$

we have

$$\begin{vmatrix} 2 & 3 \\ -4 & 7 \end{vmatrix} = 14 - (-12) = 26$$

The determinant of the matrix

$$\begin{bmatrix} a_{11} & a_{12} & a_{13} \\ a_{21} & a_{22} & a_{23} \\ a_{31} & a_{32} & a_{33} \end{bmatrix}$$

is the number

$$a_{11}\begin{vmatrix} a_{22} & a_{23} \\ a_{32} & a_{33} \end{vmatrix} - a_{12}\begin{vmatrix} a_{21} & a_{23} \\ a_{31} & a_{33} \end{vmatrix} + a_{13}\begin{vmatrix} a_{21} & a_{22} \\ a_{31} & a_{32} \end{vmatrix}$$

Note that in each term of the above sum an element of the first row is multiplied by the determinant of the matrix that would be obtained by deleting the row and the column occupied by that element. Note that the middle term, a_{12}, is preceded by a minus sign. For example,

$$\begin{vmatrix} 2 & 4 & -5 \\ 7 & 0 & 2 \\ -3 & 6 & 8 \end{vmatrix} = 2\begin{vmatrix} 0 & 2 \\ 6 & 8 \end{vmatrix} - 4\begin{vmatrix} 7 & 2 \\ -3 & 8 \end{vmatrix} + (-5)\begin{vmatrix} 7 & 0 \\ -3 & 6 \end{vmatrix}$$

$$= 2(0 - 12) - 4(56 + 6) - 5(42 - 0) = -482$$

We shall now list several well-known properties of determinants. Some of these are fairly obvious and can easily be verified by the student. Others require more complex proof.

In the interest of brevity, we point out that in each of the following statements the word "row" may be replaced by the word "column" without altering the truth of the statement. Also, for illustration purposes, determinants of 2×2 matrices will be used, although the results are true for determinants of square matrices of any size.

A constant may be factored out of any row of a determinant. Hence, if k is a scalar, then

$$\begin{vmatrix} ka_{11} & ka_{12} \\ a_{21} & a_{22} \end{vmatrix} = k \begin{vmatrix} a_{11} & a_{12} \\ a_{21} & a_{22} \end{vmatrix}$$

If two rows of a determinant are interchanged, the sign of the determinant is changed, but its absolute value is unchanged. Hence

$$\begin{vmatrix} a_{21} & a_{22} \\ a_{11} & a_{12} \end{vmatrix} = - \begin{vmatrix} a_{11} & a_{12} \\ a_{21} & a_{22} \end{vmatrix}$$

If a row of a determinant is replaced by the sum of that row and a multiple of another row, the value of the determinant is unchanged. Hence, if k is a scalar, then

$$\begin{vmatrix} a_{11} & a_{12} \\ a_{21} & a_{22} \end{vmatrix} = \begin{vmatrix} a_{11} & a_{12} \\ a_{21} + ka_{11} & a_{22} + ka_{12} \end{vmatrix}$$

If any row of a determinant consists entirely of 0's, the value of the determinant is 0. Hence the value of any determinant in which one row is a multiple of another row is 0, since the preceding property can be used to obtain a row of 0's.

The following example demonstrates the use of some of the above properties in the evaluation of a determinant.

Example 1 Find the value of the determinant

$$\begin{vmatrix} 4 & 2 & -8 \\ 3 & 6 & 12 \\ 8 & -2 & 18 \end{vmatrix}$$

Solution Factoring a 3 out of the second row and a 2 out of the third, we obtain

$$6 \begin{vmatrix} 4 & 2 & -8 \\ 1 & 2 & 4 \\ 4 & -1 & 9 \end{vmatrix}$$

Next we multiply the second row by -1 and add it to the first row, obtaining

$$6\begin{vmatrix} 3 & 0 & -12 \\ 1 & 2 & 4 \\ 4 & -1 & 9 \end{vmatrix}$$

Next we multiply the first column by 4 and add it to the third column, obtaining

$$6\begin{vmatrix} 3 & 0 & 0 \\ 1 & 2 & 8 \\ 4 & -1 & 25 \end{vmatrix}$$

Finally, we use the definition of the determinant of a 3×3 matrix to obtain

$$(6)(3)\begin{vmatrix} 2 & 8 \\ -1 & 25 \end{vmatrix} = (18)(58) = 1044$$

The definition of determinants of matrices of size greater than 3×3 is similar to the 3×3 case. For example, for the 4×4 case we have

$$\begin{vmatrix} a_{11} & a_{12} & a_{13} & a_{14} \\ a_{21} & a_{22} & a_{23} & a_{24} \\ a_{31} & a_{32} & a_{33} & a_{34} \\ a_{41} & a_{42} & a_{43} & a_{44} \end{vmatrix} = a_{11}\begin{vmatrix} a_{22} & a_{23} & a_{24} \\ a_{32} & a_{33} & a_{34} \\ a_{42} & a_{43} & a_{44} \end{vmatrix}$$

$$- a_{12}\begin{vmatrix} a_{21} & a_{23} & a_{24} \\ a_{31} & a_{33} & a_{34} \\ a_{41} & a_{43} & a_{44} \end{vmatrix} + a_{13}\begin{vmatrix} a_{21} & a_{22} & a_{24} \\ a_{31} & a_{32} & a_{34} \\ a_{41} & a_{42} & a_{44} \end{vmatrix}$$

$$- a_{14}\begin{vmatrix} a_{21} & a_{22} & a_{23} \\ a_{31} & a_{32} & a_{33} \\ a_{41} & a_{42} & a_{43} \end{vmatrix}$$

It can be seen that each term is the product of an element of the first row and the determinant of the matrix obtained by deleting the row and the column in which the element lies. Furthermore, the sign of the term is positive or negative according as the sum of the number of the row and the column in which the element lies is even or odd. This definition is general, applying to determinants of matrices of any size.

8.6 Use of Matrices and Determinants in Solving Systems of Linear Equations

Consider the following system of three linear equations:

$$x + y - z = -5 \tag{1}$$

$$2x - 3y + z = 12 \tag{2}$$
$$3x + y - 4z = -15 \tag{3}$$

Employing previously discussed operations, we can solve this system by transforming it into a sequence of equivalent systems as follows. We multiply Equation (1) by -3 and add it to Equation (3) to produce the system

$$x + y - z = -5 \tag{1}$$
$$2x - 3y + z = 12 \tag{2}$$
$$-2y - z = 0 \tag{3'}$$

We multiply Equation (1) by -2 and add it to Equation (2) to produce the system

$$x + y - z = -5 \tag{1}$$
$$-5y + 3z = 22 \tag{2'}$$
$$-2y - z = 0 \tag{3'}$$

We multiply Equation (2′) by 2 and Equation (3′) by -5 to produce the system

$$x + y - z = -5 \tag{1}$$
$$-10y + 6z = 44 \tag{2''}$$
$$10y + 5z = 0 \tag{3''}$$

We add Equation (2″) to Equation (3″) to obtain the system

$$x + y - z = -5 \tag{1}$$
$$-10y + 6z = 44 \tag{2''}$$
$$11z = 44 \tag{3'''}$$

The remainder of the solution process may now be readily carried out. From Equation (3‴) we obtain

$$z = 4$$

Substituting this value of z in Equation (2″), we obtain

$$-10y + 24 = 44$$
$$-10y = 20$$
$$y = -2$$

Substituting the values obtained for y and z in Equation (1), we get

$$x - 2 - 4 = -5$$
$$x = 1$$

As noted, the procedure used above is not new. However, we observe at this point that the process could be abbreviated by omitting the unknowns and working only with the coefficients and constant terms of the equations. If we write these down in the order in which they appear, we obtain the following array which may be written as a matrix:

$$\left[\begin{array}{rrr|r} 1 & 1 & -1 & -5 \\ 2 & -3 & 1 & 12 \\ 3 & 1 & -4 & -15 \end{array}\right]$$

The matrix consisting of the first three columns of this array is called the **coefficient matrix,** and the matrix obtained by the addition of the last column is called the **augmented matrix.** The dotted line emphasizes the fact that there are two matrices under discussion.

The steps used in the process of solving the system of equations with which the above matrix is associated could have been carried out using only the augmented matrix, and the description of the steps above would have to be modified only by replacing the word "equation" in each case by the word "row" and the word "system" by the word "matrix." For example, the first manipulation would be described as follows:

Multiply row 1 by -3 and add it to row 3 to produce the matrix

$$\left[\begin{array}{rrr|r} 1 & 1 & -1 & -5 \\ 2 & -3 & 1 & 12 \\ 0 & -2 & -1 & 0 \end{array}\right]$$

In this way the original augmented matrix can be transformed to the matrix

$$\left[\begin{array}{rrr|r} 1 & 1 & -1 & -5 \\ 0 & -10 & 6 & 44 \\ 0 & 0 & 11 & 44 \end{array}\right]$$

If the matrix is then replaced by the corresponding system of equations, the solution can then be readily completed.

Note that the operations performed on the rows are directed toward transforming the coefficient matrix to one of the form

$$\begin{bmatrix} a_{11} & a_{12} & a_{13} \\ 0 & a_{22} & a_{23} \\ 0 & 0 & a_{33} \end{bmatrix}$$

In other words, the operations are directed to transform the coefficient matrix to one with only 0's below the principal diagonal. This matrix is in **echelon** form. In general, a matrix is said to be in echelon form if $a_{ij} = 0$ for $i > j$.

The operations that may be used to transform the matrix of a system of equations to a matrix of an equivalent system may be listed as follows.

(1) Interchange any two rows.

(2) Multiply any row by a nonzero number.

(3) Add a multiple of a given row to another row.

It should be noted that in the above example additional operations of the type listed above might have been employed to further transform the last matrix in the sequence, i.e., the matrix in which the coefficient matrix was in echelon form, to the matrix

$$\left[\begin{array}{ccc|r} 1 & 0 & 0 & 1 \\ 0 & 1 & 0 & -2 \\ 0 & 0 & 1 & 4 \end{array}\right]$$

The solution set is now apparent by inspection of the matrix. However, the number of additional operations needed to produce this last matrix seems to involve more labor than was expended by stopping with the matrix in echelon form and proceeding as before.

Finally, it should be noted that matrix methods may be employed for the solution of systems in which the number of equations is not the same as the number of unknowns. Consider the following system:

$$\begin{aligned} x + y + 2z - t &= 7 \\ 2x + 8y + 6z - 8t &= -6 \\ x + y + 10z - 10t &= 11 \end{aligned}$$

The student may verify that the augmented matrix of the above system may be reduced to the echelon form shown below.

$$\begin{bmatrix} 1 & 1 & 2 & -1 & 7 \\ 0 & 6 & 2 & -6 & -20 \\ 0 & 0 & 8 & -9 & 4 \end{bmatrix}$$

This matrix corresponds to the following system, which is equivalent to the original system:

$$\begin{aligned} x+y+2z-t&=7 \\ 6y+2z-6t&=-20 \\ 8z-9t&=4 \end{aligned}$$

From the last of these equations we readily obtain

$$z=\frac{9t+4}{8}$$

If this expression for z is substituted in the second equation, we obtain

$$y=\frac{5t-28}{8}$$

Finally substituting for y and z in the first equation, we obtain

$$x=\frac{-15t+76}{8}$$

Hence, the solution set for the original system of equations may be written in the form

$$\left(\frac{-15t+76}{8}, \frac{5t-28}{8}, \frac{9t+4}{8}, t\right)$$

where t is any real number. For example, if we let $t=4$, we obtain the solution set

$$[2,-1,5,4]$$

Clearly by varying t we could generate an infinite set of solution sets.

Determinants may be used in the solution of systems of linear equations in which the number of equations is the same as the number of unknowns. The process by which the solution is carried out is known as **Cramer's rule.**

Consider the following system of two equations in two unknowns:

$$a_{11}x_1 + a_{12}x_2 = c_1$$
$$a_{12}x_1 + a_{22}x_2 = c_2$$

We now define three determinants as follows:

$$A = \begin{vmatrix} a_{11} & a_{12} \\ a_{12} & a_{22} \end{vmatrix} \qquad A_1 = \begin{vmatrix} c_1 & a_{12} \\ c_2 & a_{22} \end{vmatrix} \qquad A_2 = \begin{vmatrix} a_{11} & c_1 \\ a_{12} & c_2 \end{vmatrix}$$

Then, according to Cramer's rule:

$$x = \frac{A_1}{A} \qquad y = \frac{A_2}{A}$$

provided $A \neq 0$. Note that the determinant A is simply the determinant of the coefficients of x_1 and x_2 written in the order in which they appear in the given system of equations, whereas A_1 and A_2 are obtained by replacing the first and second columns, respectively, in A by the constant term column.

For example, consider the system

$$2x - 3y = 13$$
$$4x + 5y = -7$$

Here we have

$$A = \begin{vmatrix} 2 & -3 \\ 4 & 5 \end{vmatrix} = 22$$
$$A_1 = \begin{vmatrix} 13 & -3 \\ -7 & 5 \end{vmatrix} = 44$$
$$A_2 = \begin{vmatrix} 2 & 13 \\ 4 & -7 \end{vmatrix} = -66$$

Hence $$x = \frac{44}{22} = 2 \qquad y = \frac{-66}{22} = -3$$

For a system of three equations in three unknowns, the application of Cramer's rule follows exactly the same pattern as in the previous case. The determinant A is the determinant of the coefficients; the determinants A_1, A_2, and A_3 are obtained by replacing the first, second, and third columns, respectively, in A by the constant term column.

For example, consider the system

$$x - 4y + 3z = -24$$
$$3x + y - 2z = 13$$
$$5x - 2y + z = -6$$

Here we have

$$A = \begin{vmatrix} 1 & -4 & 3 \\ 3 & 1 & -2 \\ 5 & -2 & 1 \end{vmatrix} = 16$$

$$A_1 = \begin{vmatrix} -24 & -4 & 3 \\ 13 & 1 & -2 \\ -6 & -2 & 1 \end{vmatrix} = 16$$

$$A_2 = \begin{vmatrix} 1 & -24 & 3 \\ 3 & 13 & -2 \\ 5 & -6 & 1 \end{vmatrix} = 64$$

$$A_3 = \begin{vmatrix} 1 & -4 & -24 \\ 3 & 1 & 13 \\ 5 & -2 & -6 \end{vmatrix} = -48$$

Hence

$$x = \frac{A_1}{A} = \frac{16}{16} = 1$$
$$y = \frac{A_2}{A} = \frac{64}{16} = 4$$
$$z = \frac{A_3}{A} = \frac{-48}{16} = -3$$

If, in attempting to solve a system of equations using Cramer's rule, we find that A, the determinant of the coefficient matrix, is 0, then it may be concluded that the system of equations has no unique solution or there are an infinite number of solutions.

8.7 Markŏv Chains

A **Markŏv chain** is a sequence of experiments, each of which has a finite number of possible outcomes, and in which the outcome of a specified experiment is dependent only on the outcome of the previous experiment. The outcomes are referred to as the **states** of the Markŏv chain and will be designated as $E_1, E_2, \ldots, E_n$. The probability that at a given stage of the sequence of experiments the state E_i will be succeeded by the state E_j will be designated by p_{ij}. These probabilities may be conveniently exhibited as the elements of a square matrix known as a **transition matrix.** For example, if a Markŏv chain has three possible states, then the transition matrix may be designated as

$$P = \begin{bmatrix} p_{11} & p_{12} & p_{13} \\ p_{21} & p_{22} & p_{23} \\ p_{31} & p_{32} & p_{33} \end{bmatrix}$$

Clearly the sum of the elements in each row of a transition matrix is 1.

Furthermore, the probability that a Markŏv sequence that starts at state E_i will, after k stages, be in state E_j, will be designated by $p_{ij}^{(k)}$, and the matrix of these probabilities for the three-state Markŏv chain will be designated as

$$P^{(k)} = \begin{bmatrix} p_{11}^{(k)} & p_{12}^{(k)} & p_{13}^{(k)} \\ p_{21}^{(k)} & p_{22}^{(k)} & p_{23}^{(k)} \\ p_{31}^{(k)} & p_{32}^{(k)} & p_{33}^{(k)} \end{bmatrix}$$

Let us suppose that the probability that a given Markŏv chain starts in state E_j is denoted by $a_j^{(0)}$. Then, for the three-state Markŏv chain the probability of starting in each state will be designated by the vector $\mathbf{A}^{(0)} = (a_1^{(0)}, a_2^{(0)}, a_3^{(0)})$. A vector such as $\mathbf{A}^{(0)}$, for which the sum of the components is 1, is called a **probability vector.**

By use of the multiplication and addition laws of probability, the probability that a three-state Markŏv chain will, after one stage, arrive at state E_1 is given by

$$a_1^{(0)}p_{11} + a_2^{(0)}p_{21} + a_3^{(0)}p_{31}$$

But this probability is the first element of the product of the vector $\mathbf{A}^{(0)}$ and the matrix P, since

$$\mathbf{A}^{(0)}P = (a_1^{(0)}, a_2^{(0)}, a_3^{(0)}) \begin{bmatrix} p_{11} & p_{12} & p_{13} \\ p_{21} & p_{22} & p_{23} \\ p_{31} & p_{32} & p_{33} \end{bmatrix}$$

$$= (a_1^{(0)}p_{11} + a_2^{(0)}p_{21} + a_3^{(0)}p_{31},\ a_1^{(0)}p_{12} + a_2^{(0)}p_{22} + a_3^{(0)}p_{32},\ a_1^{(0)}p_{13} + a_2^{(0)}p_{23} + a_3^{(0)}p_{33})$$

Furthermore the second and third elements of the product $\mathbf{A}^{(0)}P$ are readily seen to be the respective probabilites that the Markŏv chain is in states E_2 and E_3 after one stage.

If we designate this product vector as

$$\mathbf{A}^{(1)} = (a_1^{(1)}, a_2^{(1)}, a_3^{(1)})$$

we have then

$$\mathbf{A}^{(1)} = \mathbf{A}^{(0)} P$$

Moreover, we readily conclude, using the same type of reasoning, that

$$\mathbf{A}^{(2)} = \mathbf{A}^{(1)} P$$

$$\mathbf{A}^{(3)} = \mathbf{A}^{(2)} P$$

and in general

$$\mathbf{A}^{(k)} = \mathbf{A}^{(k-1)} P$$

Now, by successive substitution, it is readily seen that

$$\mathbf{A}^{(2)} = \mathbf{A}^{(1)} P = (\mathbf{A}^{(0)} P) P = \mathbf{A}^{(0)} P^2$$

$$A^{(3)} = \mathbf{A}^{(2)} P = (\mathbf{A}^{(0)} P^2) P = \mathbf{A}^{(0)} P^3$$

and in general

$$\mathbf{A}^{(k)} = \mathbf{A}^{(0)} P^k$$

Example 1 In a certain family, the eldest son always goes to Harvard or Yale. If the father went to Harvard, the probabilities that the eldest son will attend Harvard and Yale are $\frac{2}{3}$ and $\frac{1}{3}$, respectively. If the father went to Yale, the probabilities that the eldest son will attend Yale and Harvard are $\frac{3}{4}$ and $\frac{1}{4}$, respectively. If the father has gone to Harvard, what are the probabilities that a grandchild who is the eldest son of his eldest son will attend Harvard and Yale, respectively?

Solution The transition matrix for this chain is

$$P = \begin{array}{c} \\ \text{H} \\ \text{Y} \end{array} \begin{array}{c} \begin{array}{cc} \text{H} & \text{Y} \end{array} \\ \begin{bmatrix} \frac{2}{3} & \frac{1}{3} \\ \frac{1}{4} & \frac{3}{4} \end{bmatrix} \end{array}$$

Also, $\mathbf{A}^{(0)} = (1,0)$

Hence, the desired probabilities are given by the vector

$$\mathbf{A}^{(2)} = \mathbf{A}^{(0)} P^2 = (1,0) \begin{bmatrix} \frac{2}{3} & \frac{1}{3} \\ \frac{1}{4} & \frac{3}{4} \end{bmatrix} \begin{bmatrix} \frac{2}{3} & \frac{1}{3} \\ \frac{1}{4} & \frac{3}{4} \end{bmatrix}$$

$$= (1,0)\begin{bmatrix} \frac{19}{36} & \frac{17}{36} \\ \frac{17}{48} & \frac{31}{48} \end{bmatrix}$$

$$\begin{matrix} \text{H} & \text{Y} \end{matrix}$$
$$= \begin{bmatrix} \frac{19}{36} & \frac{17}{36} \end{bmatrix}$$

We have seen that $\mathbf{A}^{(k)} = \mathbf{A}^{(0)}P^k$. For a three-state Markŏv chain, let us assume that $\mathbf{A}^{(0)} = (1,0,0)$. Then we have

$$\mathbf{A}^{(k)} = (1,0,0)P^k$$

But this vector is evidently the first row of the matrix P^k. Hence the elements of the first row of P^k are the probabilities that a Markŏv chain that starts in state E_1, will, after k stages, be in states E_1, E_2, and E_3, respectively. But this shows that the first row of P^k is identical to the first row of $P^{(k)}$. In the same way, using the vectors $\mathbf{A}^{(0)} = (0,1,0)$ and $\mathbf{A}^{(0)} = (0,0,1)$, we readily see that the second and third rows of P^k are also identical to the second and third rows of $P^{(k)}$. Hence we have

$$P^{(k)} = P^k$$

A probability vector $\mathbf{A}$ is called a **fixed point** of a transition matrix P if

$$\mathbf{A}P = \mathbf{A}$$

A transition matrix is called regular if all of the elements of P^k are positive for some positive integer k. The matrix P of Example 1 is regular since, as we have seen, the elements of P^2 are all positive.

It can be proved that the fixed point of a regular transition matrix is unique.

Example 2 Find the fixed point of the matrix P of Example 1.
Solution If the fixed point is designated as (a_1,a_2), then

$$(a_1,a_2)\begin{bmatrix} \frac{2}{3} & \frac{1}{3} \\ \frac{1}{4} & \frac{3}{4} \end{bmatrix} = (a_1,a_2)$$

Hence, a_1 and a_2 must satisfy the equations

$$a_1 + a_2 = 1$$

$$\frac{2}{3}a_1 + \frac{1}{4}a_2 = a_1$$

$$\frac{1}{3}a_1 + \frac{3}{4}a_2 = a_2$$

where the first equation follows from the fact that (a_1,a_2) is a probability vector.

Solving the system of equations, we obtain the vector $(\frac{3}{7},\frac{4}{7})$ as the fixed point of the given transition matrix.

The importance of the fixed point of a specified transition matrix is accentuated by the following remarkable fact. It can be proved that if P is a regular transition matrix, then as k increases, the matrix P^k approaches a matrix, each row of which is the fixed point of P. Stated more formally,

$$\lim_{k\to\infty} P^k = B$$

where each row of B is the fixed point A of P.

Hence, for the transition matrix of Example 1, we have

$$\lim_{k\to\infty} P^{(k)} = \lim_{k\to\infty} P^k = \begin{bmatrix} \frac{3}{7} & \frac{4}{7} \\ \frac{3}{7} & \frac{4}{7} \end{bmatrix}$$

We may interpret this result to mean that regardless of whether a father has attended Harvard or Yale and assuming a line of eldest sons, the probabilities that after many generations an eldest son will attend Harvard and Yale will approach $\frac{3}{7}$ and $\frac{4}{7}$, respectively.

Problems—Chapter 8

1. Given the vectors

$$\mathbf{A} = (2,-5,3), \quad \mathbf{B} = (-1,4,6)$$

find: (a) $\mathbf{A} - \mathbf{B}$ (c) $\mathbf{A} \cdot \mathbf{B}$
(b) $3\mathbf{A} + \mathbf{B}$ (d) $||\mathbf{A} + \mathbf{B}||$

2. A force of 30 lb acts on an object in the horizontal direction. A second force of 20 lb acts on the object at an angle of 35° to the horizontal. Find the magnitude and the direction of the resultant force.

3. Given the vectors **A** = (2,4,−3) and **B** = (6,3,x), determine x so that the vectors will be perpendicular to each other.
4. Find to the nearest degree the angle between the vectors **A** = (3,2,1) and **B** = (4,5,−3).
5. Given the matrices

$$A=\begin{bmatrix}2 & 4\\ 3 & -6\end{bmatrix} \quad \text{and} \quad B=\begin{bmatrix}-4 & 5\\ 8 & 2\end{bmatrix}$$

find: (a) $A+B$ (c) AB
(b) $3A$ (d) A^{-1}

6. Find the determinant of the matrix:

$$A=\begin{bmatrix}1 & 6 & 3\\ 2 & -1 & 5\\ 4 & 2 & -3\end{bmatrix}$$

7. Given the system of equations

$$\begin{aligned} x+y+z&=0\\ 2x-3y+z&=-2\\ 4x-y+2z&=1 \end{aligned}$$

(a) Obtain the augmented matrix of the system.
(b) Transform this matrix to echelon form.
(c) Use the echelon form of the matrix to obtain the solution of the system.

8. For the system of equations given in Problem 7, obtain the solution using determinants (Cramer's rule).
9. In a certain locality, if it rains on a given day, the probability of rain on the next day is $\frac{3}{4}$. If it does not rain on a given day, the probability of rain the next day is $\frac{1}{5}$.
(a) Thinking of this as a two-state Markŏv chain, obtain the transition matrix P.
(b) Obtain a matrix that exhibits the probabilities of rain or no rain on a specified day, given the state of rain or no rain two days previously.
(c) If it rains on a given Monday, what is the probability that it will not rain on Wednesday of the same week?
(d) What is the probability that it will rain on a given day after a very long period of time?

Answers to Problems—Chapter 1

1. **(a)** $\{1,5,9,11,13,15\}$
(b) $\{5,9\}$
(c) $\{1,11\}$
(d) $\{3,7,13,15\}$
(e) $\{1,3,7,11,13,15\}$
(f) $\{1,3,7,11,13,15\}$
(g) $\{3,7\}$
(h) $\{3,7\}$
(i) $\{(1,5), (1,9), (1,13), (1,15), (5,5), (5,9), (5,13), (5,15), (9,5), (9,9), (9,13), (9,15), (11,5) (11,9), (11,13), (11,15)\}$
(j) $\{(5,1), (5,5), (5,9), (5,11), (9,1), (9,5), (9,9), (9,11), (13,1), (13,5), (13,9), (13,11), (15,1), (15,5), (15,9), (15,11)\}$

2. **(a)** 32
(b) 6
(c) 33
(d) 37
(e) 13
(f) 44
(g) 19
(h) 5
(i) 7

3. **(a)** $\dfrac{y^{12}}{x^8}$
(b) $\dfrac{y^2 z}{x^2}$
(c) $\dfrac{1}{x^{14/3}y^4}$

4. **(a)** $\dfrac{1}{125}$
(b) $\dfrac{9}{4}$
(c) 3125
(d) $\dfrac{1}{128}$

5. **(a)** $6\sqrt[3]{7}$
(b) $2xy^2\sqrt[4]{2x^2}$
(c) $-5x\sqrt{2}$
(d) $\sqrt[15]{x^8}$
(e) $\sqrt[4]{x}$

6. **(a)** $(4x^2 + 9y^2)(2x + 3y)(2x - 3y)$
(b) $(x^2 + 4y^2)(x^4 - 4x^2y^2 + 16y^2)$
(c) $(x^5 - y^4)(x^{10} + x^5y^4 + y^8)$
(d) $(x - y)(c + d)$
(e) $(a + b)(x - y)(x^2 + xy + y^2)$

7. **(a)** $y = -3x$
(b) $y = \dfrac{3}{16x}$
(c) $q = \dfrac{8xy}{z}$

8. (a) $5 - i$
(b) $26 - 7i$
(c) $-\frac{14}{29} + \frac{23}{29}i$
(d) $\frac{7}{50} + \frac{1}{50}i$
(e) $2 + 11i$

9. (a) 2
(b) ± 4
(c) $-1, 5$
(d) $-4, 0, 7$
(e) 5, 5
(f) $-2 \pm \sqrt{2}$
(g) $-3 \pm i$

10. (a) $x = -2, y = 6$
(b) No solution
(c) $x = 4, y = 3, z = -2$

11. (a) $(-\infty, 6)$
(b) $[-6, \infty)$
(c) $(-10, 4)$
(d) $(-\infty, -\frac{11}{5}] \cup [3, \infty)$
(e) $(2, 5)$
(f) $(-\infty, \infty)$

12. 5040

13. $\frac{2}{0}$

14. 126

15. (a) 120
(b) 36

16. (a) 10
(b) 38
(c) 110
(d) 44
(e) 190

Answers to Problems—Chapter 2

1.

	Domain	**Range**
(a)	$\{2,3,4\}$	$\{5,7,9\}$
(b)	$[-\infty,7]$	$[0,\infty)$
(c)	$[-4,4]$	$(-\infty,0]$
(d)	$(-\infty,-3) \cup (-3,\infty)$	$(-\infty,0) \cup (0,\infty)$
(e)	$(-\infty,2) \cup (2,\infty)$	$(0,\infty)$
(f)	$(-\infty,-5) \cup (5,\infty)$	$(0,\infty)$

2. **(a)** -4

(b) 4

(c) $3p^2 - 2p - 4$

(d) 0

(e) $3x + 2\sqrt{x+4+8}$

(f) $\sqrt{3x^2 + 2x}$

3. $C = \begin{cases} 10x & x = 0, 1, 2, 3, \ldots, 30 \\ 7x + 90 & x = 31, 32, \ldots \end{cases}$

4. **(a)**

x	y
0	-4
± 1	-3
± 2	0
± 3	5

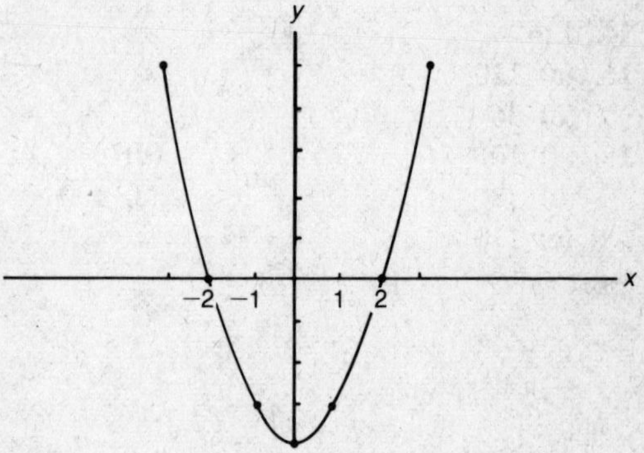

$y = x^2 - 4$

(b)

x	y
± 4	0
± 5	3
± 6	$2\sqrt{5} \approx 4.5$

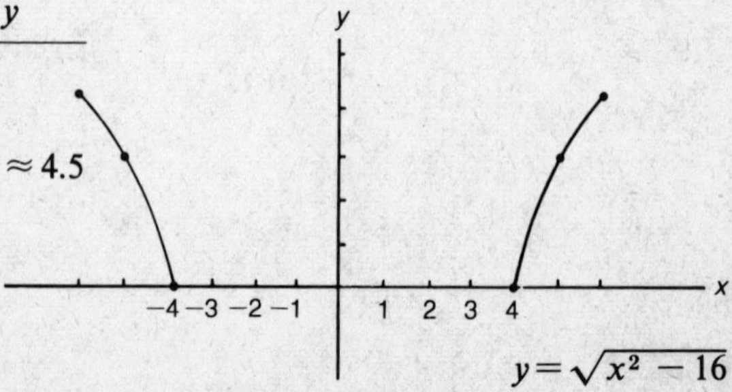

$y = \sqrt{x^2 - 16}$

(c)

x	y
-7	$-\frac{1}{4}$
-6	$-\frac{1}{3}$
-5	$-\frac{1}{2}$
-4	-1
$-\frac{13}{4}$	-4
$-\frac{11}{4}$	4
-2	1
-1	$\frac{1}{2}$
0	$\frac{1}{3}$
1	$\frac{1}{4}$

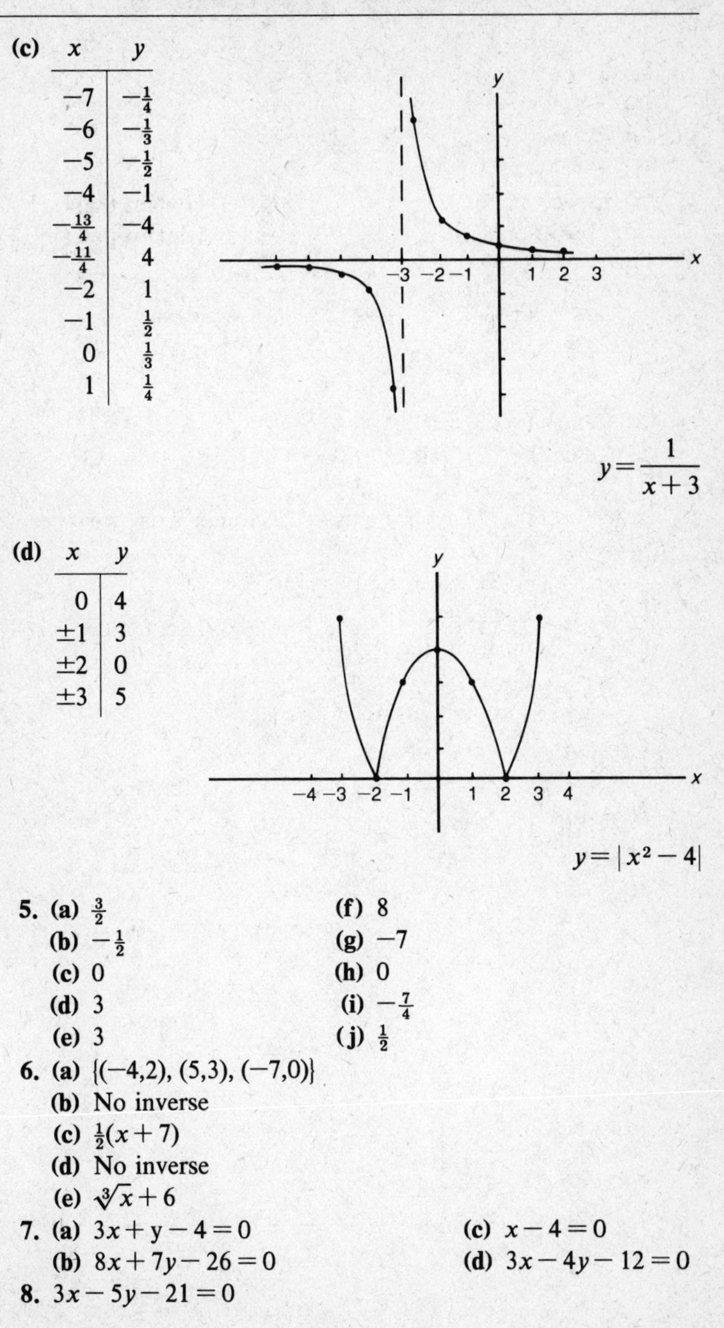

$$y = \frac{1}{x+3}$$

(d)

x	y
0	4
± 1	3
± 2	0
± 3	5

$$y = |x^2 - 4|$$

5. **(a)** $\frac{3}{2}$
(b) $-\frac{1}{2}$
(c) 0
(d) 3
(e) 3
(f) 8
(g) -7
(h) 0
(i) $-\frac{7}{4}$
(j) $\frac{1}{2}$

6. **(a)** $\{(-4,2), (5,3), (-7,0)\}$
(b) No inverse
(c) $\frac{1}{2}(x+7)$
(d) No inverse
(e) $\sqrt[3]{x}+6$

7. **(a)** $3x + y - 4 = 0$
(b) $8x + 7y - 26 = 0$
(c) $x - 4 = 0$
(d) $3x - 4y - 12 = 0$

8. $3x - 5y - 21 = 0$

9. $3x+y+11=0$

10. **(a)** 8 **(c)** $\sqrt{p^2+4q^2}$
(b) 5

11. **(a)** (2,7) **(c)** $(\frac{1}{2},\frac{27}{4})$
(b) (0,6)

12. **(a)** Lowest point: (2,−7) **(c)** Highest point: (1,1)
(b) Lowest point: (−4,−26) **(d)** Highest point: $(-\frac{1}{2},\frac{13}{4})$

13. **(a)** 125 **(d)** $\frac{16}{5}$
(b) $\frac{1}{2}$ **(e)** $\dfrac{\log 42}{\log 4} \simeq 2.70$
(c) 2

14. **(a)** Center (−2,7), radius = 4
(b) Center (−5,2), radius = 0
(c) Center $(2,\frac{3}{2})$, radius = 3

15. Center = (1 − 3), major axis = 12, minor axis = 4, eccentricity = $2\sqrt{2}/3$

16. Center = (−5,3), eccentricity = $\frac{5}{3}$

17. **(a)** (4,−3), opens upward
(b) (2,−5), opens to the right
(c) (0,−7), opens downward

18. **(a)** $x=-3$, $y=1$ $\quad$ $x=-39/19$, $y=-35/19$
(b) $x=-3$, $y=6$ $\quad$ $x=-2$, $y=6$
(c) $x=4$, $y=3$ $\quad$ $x=-4$, $y=3$ $\quad$ $x=4$, $y=-3$ $\quad$ $x=-4$, $y=-3$

19.

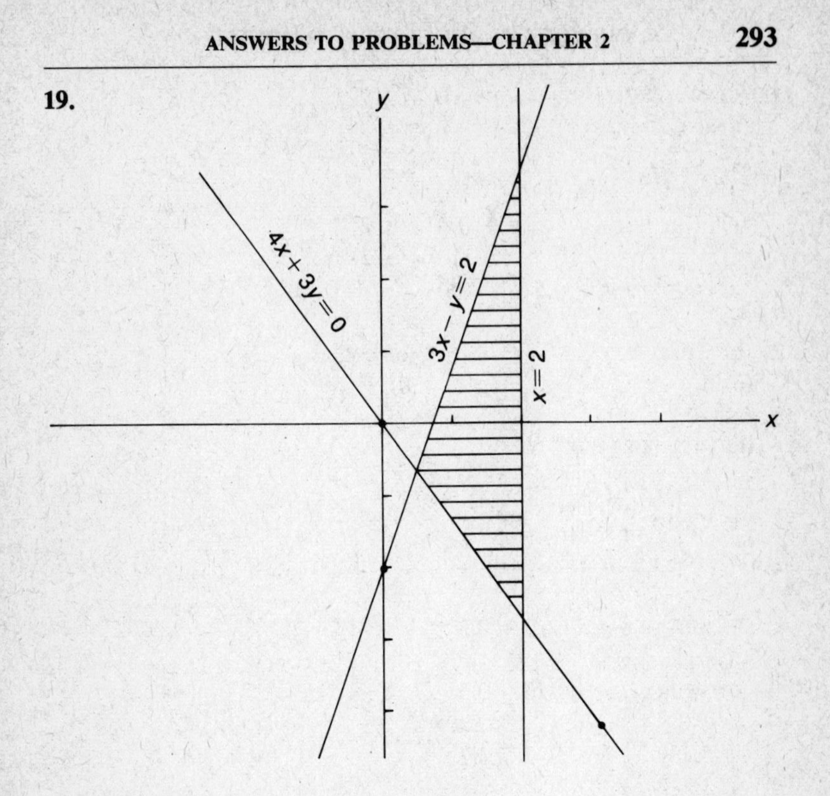

20. Maximum = 28, minimum = 4/13

21. Let x = number of job A workers

y = number of job B workers

Then $0 \leq x \leq 40$

$0 \leq y \leq 30$

$3x + 4y \leq 192$

The function $x + y$ is maximized when $x = 40$ and $y = 18$.

Answers to Problems—Chapter 3

1. **(a)** $-\sqrt{3}/2$
(b) $-\sqrt{3}$
(c) $\sqrt{2}/2$
(d) $-\frac{1}{2}$
(e) $-\frac{1}{2}$
(f) 1
(g) $-\sqrt{3}/3$
(h) $-\sqrt{3}/2$
(i) $\sqrt{2}/2$
(j) $\frac{1}{2}$
(k) $\sqrt{3}/2$
(l) 1
(m) $\sqrt{3}/2$
(n) $\sqrt{3}/2$
(o) $\sqrt{3}$

2. **(a)** $2\csc^2\alpha$
(b) 1
(c) $\cot^2\alpha$
(d) 1

3. **(a)** 120°, 240°
(b) 30°, 90°, 150°
(c) 90°, 135°, 315°
(d) 30°, 150°, 210°, 330°

4. **(a)**

θ	r
0	1
$\pi/4$	$1-\sqrt{2}/2\approx 0.3$
$\pi/2$	0
$3\pi/4$	$1-\sqrt{2}/2\approx 0.3$
π	1
$5\pi/4$	$1+\sqrt{2}/2\approx 1.7$
$3\pi/2$	2
$7\pi/4$	$1+\sqrt{2}/2\approx 1.7$

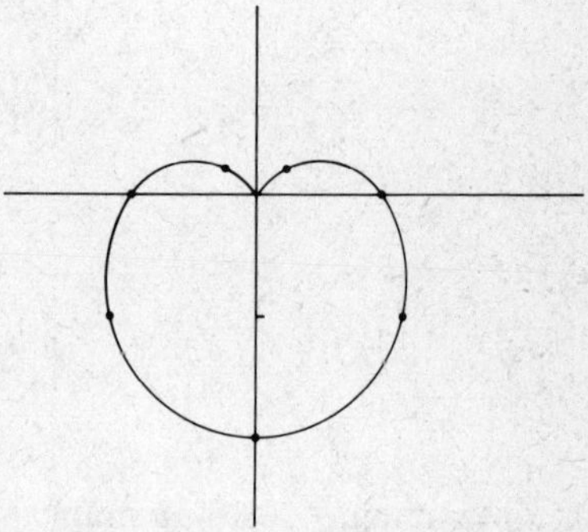

$r = 1 - \sin\theta$

(b)

θ	r
0	3
$\pi/4$	2.1
$\pi/2$	0
$3\pi/4$	-2.1
π	-3

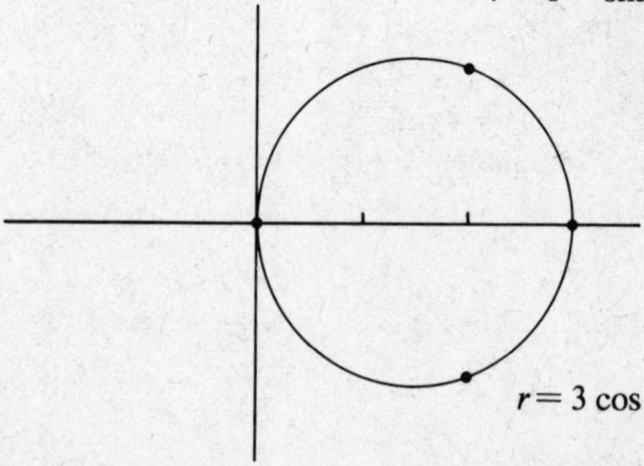

$r = 3\cos\theta$

5. $10\sqrt{2} + 10\sqrt{2}\,i$
6. $-2\sqrt{3} + 2\sqrt{3}\,i$
7. $4, -2 + 2\sqrt{3}\,i, -2 - 2\sqrt{3}\,i$

8. **(a)** $-\pi/6$
(b) $5\pi/6$
(c) 0
(d) $\pi/2$
(e) π
(f) $-\pi/6$
(g) $\pi/3$
(h) $5\pi/6$
(i) $-\pi/4$

9. Start with the right-hand side and show that it reduces to the expression on the left.

Answers to Problems—Chapter 4

1. 11/20

2. **(a)** 1/13
(b) 4/13
(c) 25/52

3. **(a)** 1/18
(b) 7/18

4.

x	2	3	4	5	6	7	8	9	10	11	12
$p(x)$	$\frac{1}{36}$	$\frac{2}{36}$	$\frac{3}{36}$	$\frac{4}{36}$	$\frac{5}{36}$	$\frac{6}{36}$	$\frac{5}{36}$	$\frac{4}{36}$	$\frac{3}{36}$	$\frac{2}{36}$	$\frac{1}{36}$

5. **(a)** 4/15
(b) 32/105
(c) 4/5
(d) 37/105
(e) 68/105

6. **(a)** 1/4
(b) 1/24
(c) 23/24
(d) 11/24

7. **(a)** 44/91
(b) 58/91

8. **(a)** 80/243
(b) 496/729

9. **(a)** 5/16
(b) 11/32

10. $\frac{(6!) \cdot (3!)}{8!} = \frac{3}{28}$

11. $E(x) = 3/4$, var $(x) = 19/16$

12. Yes. The expected value of each player's winnings is 0.

13. $(n^2 - 1)/12$

14. 5/13

Answers to Problems—Chapter 5

1. (a) $6x-2$

(b) $\dfrac{-4}{x^3}$

(c) $\dfrac{2}{(x+2)^2}$

(d) m

2. (a) $21x^2-28x+5$

(b) $5x^{3/2}-\dfrac{1}{x^{2/3}}$

(c) $\dfrac{-7}{x^2}+\dfrac{8}{x^3}$

(d) $\dfrac{1}{3x^{2/3}}+\dfrac{1}{2x^{3/2}}$

(e) $12(3x-5)^3$

(f) $\dfrac{1}{\sqrt{2x-7}}$

(g) $\dfrac{2x}{3\sqrt[3]{x^2+10}}$

(h) $-\dfrac{1}{(2x-6)^{3/2}}$

(i) $5\left(1-\dfrac{1}{x}\right)^4\left(\dfrac{1}{x^2}\right)$

3. (a) $20x^3+36x^2-12x+$

(b) $180(3x-4)^3$

(c) $-\dfrac{1}{4(x-7)^{3/2}}$

(d) $\dfrac{6}{(2x-3)^{5/2}}$

4. (a) $\dfrac{16x-5x^2}{2\sqrt{4-x}}$

(b) $-\dfrac{x^2+3}{(x^2-3)^2}$

(c) $\dfrac{3x^2-10x}{(2x-5)^{3/2}}$

5. (a) $3x^2\cos(x^3-4)$

(b) $-6\sin 3x\cos 3x=-3\sin 6x$

(c) $\dfrac{1}{2\sqrt{x}}\sec^2\sqrt{x}$

(d) $\sec^3 x+\sec x\tan^2 x$

(e) $\dfrac{3x^2-4}{x^3-4x}$

(f) $\dfrac{2}{2x-3}-\dfrac{4}{4x-5}$

(g) $e^{\sin x}\cos x$

(h) $e^x(\cos 2x-2\sin 2x)$

(i) $\dfrac{1}{2\sqrt{x}\sqrt{1-x}}$

(j) $-\dfrac{4x^3}{\sqrt{1-x^8}}$

(k) $e^x\left(\dfrac{2}{1+4x^2}+\arctan 2x\right)$

(l) $x^3\sinh x+3x^2\cosh x$

6. (a) $12x-y-18$

(b) $6\sqrt{3}x-12y+6-\sqrt{3}\pi=0$

(c) $3x-y+1=0$

7. (a) $f(-2)=\dfrac{43}{3}$ is a relative maximum

$f(4)=-\dfrac{65}{3}$ is a relative minimum

(b) $f(0)=0$ is a relative minimum
$f(2)=4e^{-2}$ is a relative maximum
(c) $f(1)=-5$ is an endpoint maximum
$f(3)=-9$ is a relative and absolute minimum
$f(6)=0$ is an endpoint and absolute maximum

8. 5, 5

9. **(a)** 4 ft to the right of the origin
(b) -3 ft/sec
(c) $0 \le t < 1$
(d) 18 ft/sec

10. **(a)** $t=6$
(b) 160 ft/sec
(c) 400 ft

11. 1/2 unit/sec

12. 240 cm^3/sec

13. **(a)** -1
(b) 1
(c) 3

14. **(a)** $\frac{1}{6}x^6+\frac{1}{2}x^4-\frac{5}{2}x^2+7x+C$
(b) $\dfrac{2\sqrt{2}}{3}x^{3/2}+C$
(c) $\frac{3}{4}x^{4/3}+C$
(d) $-\dfrac{1}{x}-\dfrac{1}{x^2}+C$
(e) $2x^{1/2}+\frac{8}{3}x^{3/4}+C$
(f) $\frac{1}{3}x^3-\frac{4}{5}x^{5/2}+\frac{1}{2}x^2+C$
(g) $2x^4-4x^3+3x^2-x+C$ or $\frac{1}{8}(2x-1)^4+C$
(h) $\frac{2}{5}x^{5/2}-\frac{2}{3}x^{3/2}-4x^{1/2}+C$
(i) $\frac{1}{18}(2x-7)^9+C$
(j) $-\frac{1}{15}(5-3x)^5+C$
(k) $\frac{1}{6}(4x-7)^{3/2}+C$
(l) $-2\sqrt{3-x}+C$
(m) $\frac{1}{3}(x^2-5)^{3/2}+C$
(n) $\frac{1}{2}(x^3-7)^{2/3}+C$
(o) $-\dfrac{1}{2(x^2-6x-1)}+C$

15. $Y=\frac{1}{3}x^3-5$

16. $A=\frac{1}{3}t^3-\frac{2}{3}t^{3/2}+4$

17. **(a)** $-\frac{1}{4}e^{-4x}+C$
(b) $-\frac{1}{3}e^{-x^3}+C$
(c) $2e^{\sqrt{x}}+C$
(d) $\frac{1}{3}\ln|3x-5|+C$
(e) $-\frac{1}{2}\ln|4-x^2|+C$
(f) $-\frac{1}{4}\cos 4x+C$
(g) $2\ln|\sec\frac{1}{2}x|+C$
(h) $\frac{1}{6}\sin^3 2x+C$
(i) $\frac{1}{5}\ln|5\sin x-4|+C$
(j) $\frac{1}{4}(\ln x)^4+C$
(k) $\frac{1}{4}(1+\tan x)^4+C$
(l) $\frac{1}{3}\arcsin 3x/4+C$
(m) $\frac{1}{10}\arctan 2x/5+C$

18. **(a)** $-\frac{1}{3}xe^{-3x}-\frac{1}{9}e^{-3x}+C$
(b) $-x\cos x+\sin x+C$
(c) $x^2e^x-2xe^x+2e^x+C$
(d) $\frac{1}{2}x^2\ln x-\frac{1}{4}x^2+C$

19. **(a)** 26/3
(b) 15/4
(c) 2/11
(d) $\frac{1}{2}(e^6-1)$
(e) 3/4
(f) $\pi/16$

20. **(a)** 64/3
(b) 1/3
(c) 36
(d) $e-e^{-2}\simeq 2.58$

21. $-41/4$

22. **(a)** 1/8
(b) Diverges
(c) 4
(d) 4

23. **(a)** $3x^2y^2+y^4$
(b) $2x^3y+4xy^3$
(c) $6x^2y+4y^3$

24. **(a)** $2xye^{x^2y}+y\cos xy$
(b) $x^2e^{x^2y}+x\cos xy$
(c) $2x^3ye^{x^2y}+2xe^{x^2y}-xy\sin xy+\cos xy$

25. **(a)** $y=\frac{1}{2}x^3-\frac{1}{3}x^4+c_1x+c_2$
(b) $y=Ce^{2x}-\frac{3}{2}$
(c) $y=C_1e^x+C_2e^{4x}$

26. **(a)** $P=40{,}000\,(\frac{9}{10})^{t/5}$
(b) 32,400

Answers to Problems—Chapter 6

1. 4

2. **(a)** 1/27 **(b)** 7/64

3. $1 - e^{-1/2}$

4. **(a)** $F(x) = \begin{cases} 0, & x \leq 0 \\ \dfrac{x^2}{4}, & 0 < x < 2 \\ 1, & x \geq 2 \end{cases}$

(b) 9/16

5. **(a)** $\mu = \frac{4}{3}$

(b) $\alpha^2 = \frac{2}{9}$

6. **(a)** 0.1359

(b) 0.0808

(c) 0.7865

(d) 0.0548

(e) 0.9641

7. 86

8. 0.7580

Answers to Problems—Chapter 7

1. (a) 4
   (b) 3
   (c) 2.4
   (d) 10
   (e) 3.16
2. 0.0668
3. [28.71, 31.29]
4. [29.18, 30.82]
5. [0.12, 0.28]
6. $\overline{x} > 30.82$
7. $\overline{x} < 29.02$ or $\overline{x} > 30.98$
8. $\beta(32 \simeq 0.01$. This is the probability that H_0 will be accepted when the true value of the mean is 32.
9. $p' < 0.73$
10. $p' < 0.72$ or $p' > 0.88$

Answers to Problems—Chapter 8

1. **(a)** $(3,-9,-3)$
(b) $(5,-11,15)$
(c) (-40)
(d) $\sqrt{83}$

2. Approximate values: magnitude = 47.8 lb, direction = 14°

3. 8

4. 44°

5. **(a)** $\begin{bmatrix} -2 & 9 \\ 11 & -4 \end{bmatrix}$

(b) $\begin{bmatrix} 6 & 12 \\ 9 & -18 \end{bmatrix}$

(c) $\begin{bmatrix} 24 & 18 \\ -60 & 3 \end{bmatrix}$

(d) $\begin{bmatrix} \frac{1}{4} & \frac{1}{6} \\ \frac{1}{8} & -\frac{1}{12} \end{bmatrix}$

6. 173

7. **(a)** $\begin{bmatrix} 1 & 1 & 1 & 0 \\ 2 & -3 & 1 & -2 \\ 4 & -1 & 2 & 1 \end{bmatrix}$

(b) $\begin{bmatrix} 1 & 1 & 1 & 0 \\ 0 & -5 & -1 & -2 \\ 0 & 0 & -1 & 3 \end{bmatrix}$

(c) $x=2,\ y=1,\ z=-3$

8. Same as 7(c).

9. **(a)** $\begin{bmatrix} \frac{3}{4} & \frac{1}{4} \\ \frac{1}{5} & \frac{4}{5} \end{bmatrix}$

(b) $\begin{bmatrix} \frac{49}{80} & \frac{31}{80} \\ \frac{31}{100} & \frac{69}{100} \end{bmatrix}$

(c) $\frac{31}{80}$

(d) $\frac{4}{9}$

APPENDIX I

STANDARD NORMAL CURVE AREAS*

$$A(z) = \int_0^z \frac{1}{\sqrt{2\pi}} e^{-\frac{1}{2}t^2}\, dt$$

Z	.00	.01	.02	.03	.04	.05	.06	.07	.08	.09
0.0	.0000	.0040	.0080	.0120	.0160	.0199	.0239	.0279	.0319	.0359
0.1	.0398	.0438	.0478	.0517	.0557	.0596	.0636	.0675	.0714	.0753
0.2	.0793	.0832	.0871	.0910	.0948	.0987	.1026	.1064	.1103	.1141
0.3	.1179	.1217	.1255	.1293	.1331	.1368	.1406	.1443	.1480	.1517
0.4	.1554	.1591	.1628	.1664	.1700	.1736	.1772	.1808	.1844	.1879
0.5	.1915	.1950	.1985	.2019	.2054	.2088	.2123	.2157	.2190	.2224
0.6	.2257	.2291	.2324	.2357	.2389	.2422	.2454	.2486	.2517	.2549
0.7	.2580	.2611	.2642	.2673	.2704	.2734	.2764	.2794	.2823	.2852
0.8	.2881	.2910	.2939	.2967	.2995	.3023	.3051	.3079	.3106	.3133
0.9	.3159	.3186	.3212	.3238	.3264	.3289	.3315	.3340	.3365	.3389

APPENDIX I—Continued

Z	.00	.01	.02	.03	.04	.05	.06	.07	.08	.09
1.0	.3413	.3438	.3461	.3485	.3508	.3531	.3554	.3577	.3599	.3621
1.1	.3643	.3665	.3686	.3708	.3729	.3749	.3770	.3790	.3810	.3830
1.2	.3849	.3869	.3888	.3907	.3925	.3944	.3962	.3980	.3997	.4015
1.3	.4032	.4049	.4066	.4082	.4099	.4115	.4131	.4147	.4162	.4177
1.4	.4192	.4207	.4222	.4236	.4251	.4265	.4279	.4292	.4306	.4319
1.5	.4332	.4345	.4357	.4370	.4382	.4394	.4406	.4418	.4429	.4441
1.6	.4452	.4463	.4474	.4484	.4495	.4505	.4515	.4525	.4535	.4545
1.7	.4554	.4564	.4573	.4582	.4591	.4599	.4608	.4616	.4625	.4633
1.8	.4641	.4649	.4656	.4664	.4671	.4678	.4686	.4693	.4699	.4706
1.9	.4713	.4719	.4726	.4732	.4738	.4744	.4750	.4756	.4761	.4767
2.0	.4773	.4778	.4783	.4788	.4793	.4798	.4803	.4808	.4812	.4817
2.1	.4821	.4826	.4830	.4834	.4838	.4842	.4846	.4850	.4854	.4857
2.2	.4861	.4864	.4868	.4871	.4875	.4878	.4881	.4884	.4887	.4890

2.3	.4893	.4896	.4898	.4901	.4904	.4906	.4909	.4911	.4913	.4916
2.4	.4918	.4920	.4922	.4925	.4927	.4929	.4931	.4932	.4934	.4936
2.5	.4938	.4940	.4941	.4943	.4945	.4946	.4948	.4949	.4951	.4952
2.6	.4953	.4955	.4956	.4957	.4959	.4960	.4961	.4962	.4963	.4964
2.7	.4965	.4966	.4967	.4968	.4969	.4970	.4971	.4972	.4973	.4974
2.8	.4974	.4975	.4976	.4977	.4977	.4978	.4979	.4979	.4980	.4981
2.9	.4981	.4982	.4983	.4983	.4984	.4984	.4985	.4985	.4986	.4986
3.0	.4987	.4987	.4987	.4988	.4988	.4989	.4989	.4989	.4990	.4990
3.1	.4990	.4991	.4991	.4991	.4992	.4992	.4992	.4992	.4993	.4993
3.2	.4993	.4993	.4994	.4994	.4994	.4994	.4994	.4995	.4995	.4995
3.3	.4995	.4995	.4996	.4996	.4996	.4996	.4996	.4996	.4996	.4997
3.4	.4997	.4997	.4997	.4997	.4997	.4997	.4997	.4997	.4997	.4998
3.5	.4998									

APPENDIX II

DERIVATIVES

NOTE: a is a constant in the formulas below.

$f(x)$	$f'(x)$
a	0
e^x	e^x
a^x	$a^x \ln a$
$\ln x$	$\frac{1}{x}$
$\log x$	$\frac{1}{x \log e}$
$\sin x$	$\cos x$
$\cos x$	$-\sin x$
$\tan x$	$\sec^2 x$
$\csc x$	$-\csc x \cot x$
$\sec x$	$\sec x \tan x$
$\cot x$	$-\csc^2 x$
$\arcsin x$	$\frac{1}{\sqrt{1-x^2}}$
$\arccos x$	$-\frac{1}{\sqrt{1-x^2}}$
arctax x	$\frac{1}{1+x^2}$
$\sinh x$	$\cosh x$
$\cosh x$	$\sinh x$
$\tanh x$	$\text{sech}^2 x$
$ag(x)$	$ag'(x)$
$g(x) + h(x)$	$g'(x) + h'(x)$
$g(x) \cdot h(x)$	$g(x)h'(x) + g'(x)h(x)$
$\frac{g(x)}{h(x)}$	$\frac{h(x)g'(x) - g(x)h'(x)}{[h(x)]^2}$
$g[h(x)]$	$g'[h(x)]h'(x)$

APPENDIX III

INTEGRALS

NOTE: The arbitrary constant C has been omitted from the integrals below. The constant a is assumed to be positive except in the first formula.

$f(x)$	$\int f(x)\,dx$
a	ax
x^n	$\frac{x^{n+1}}{n+1}$, $n \neq -1$
$\frac{1}{x}$	$\ln \lvert x \rvert$
e^x	e^x
a^x	$\frac{a^x}{\ln a}$
$\ln x$	$x \ln x - x$
$\sin x$	$-\cos x$
$\cos x$	$\sin x$
$\tan x$	$\ln \lvert \sec x \rvert$
$\csc x$	$\ln \lvert \csc x - \cot x \rvert$
$\sec x$	$\ln \lvert \sec x + \tan x \rvert$
$\cot x$	$\ln \lvert \sin x \rvert$
$\sec^2 x$	$\tan x$
$\sec x \tan x$	$\sec x$
$\csc^2 x$	$-\cot x$
$\csc x \cot x$	$-\csc x$
$\sinh x$	$\cosh x$
$\cosh x$	$\sinh x$
$\frac{1}{a^2 + x^2}$	$\frac{1}{a} \arctan \frac{x}{a}$
$\frac{1}{\sqrt{a^2 - x^2}}$	$\arcsin \frac{x}{a}$
$\frac{1}{x^2 - a^2}$	$\frac{1}{2a} \ln \left\lvert \frac{x-a}{x+a} \right\rvert$
$\frac{1}{\sqrt{x^2 \pm a^2}}$	$\ln (x + \sqrt{x^2 \pm a^2})$

$$\int f(x)g'(x)\,dx = f(x)g(x) - \int g(x)f'(x)\,dx$$

$$\int f[g(x)]g'(x)\,dx = F[g(x)],$$

where $F'(x) = f(x)$

INDEX